A. Yokota, T. Fujii, K. Goto (Eds.)

Alicyclobacillus

Thermophilic Acidophilic Bacilli

A. Yokota, T. Fujii, K. Goto (Eds.)

Alicyclobacillus

Thermophilic Acidophilic Bacilli

With 39 Figures

Springer

Akira Yokota, Ph.D.
Associate Professor, Institute of Molecular and Cellular Biosciences
The University of Tokyo

Tateo Fujii, Ph.D.
Professor, Department of Food Science
Yamawaki Gakuen Junior College
Formerly Professor, Department of Food Science and Technology
Tokyo University of Marine Science and Technology

Keiichi Goto, Ph.D.
Food Research Laboratories
Mitsui Norin Co.,Ltd.
223-1 Miyahara, Fujieda, Shizuoka 426-0133, Japan

ISBN 978-4-431-69849-4 Springer Tokyo Berlin Heidelberg New York
Library of Congress Control Number: 2007922357

Springer is a part of Springer Science+Business Media
springer.com

Typesetting: Camera-ready by the editors and authors

Printed on acid-free paper

Foreword

Soft drinks with pHs lower than 4.0 are subjected to minimum pasteurization at 65 °C for 10 min as required by the Japanese Food Sanitation Law. Not only pathogenic bacteria but most spore-forming bacteria are unable to grow at this low pH condition, and thus reports of microbial spoilage in pasteurized acidic soft drinks are rare.

Since 1982, when the spoilage of aseptically packed apple juice was attributed to a new type of acidophilic spore-forming bacteria in Germany, a succession of similar complaints regarding other fruit juice concentrates and their products has been received. In the beginning, the bacteria were classified in the genus *Bacillus*, but later, in 1992, the new genus *Alicyclobacillus* was proposed owing to their characteristic cellular membranes containing omega-alicyclic fatty acids. A group of *Alicyclobacillus* strains, responsible for the tainting of fruit juices, was then described as *A. acidoterrestris* in 1999. They are acidophilic and grow preferably at around pH 4.0. They are thermophilic and grow better at temperatures above 40 °C. This indicates that we might have been missing them by our ordinary methods of bacterial detection at pH 7.0 and 35 °C. Their spores are not inactivated by the pasteurizing conditions generally applied to juice concentrates and juice-containing beverages. Above all, because the bacteria do not produce gas, consumers do not see any sign of spoilage until they open the product and notice its unpleasant taint.

The off-flavor is generated by the strain's metabolism, which converts vanillin in the original juices and flavor ingredients into guaiacol, which has a strong medicinal, phenolic odor. *A. acidoterrestris* from soil and manufacturing environments contaminates juices, then the spores germinate and begin growing when conditions are conducive. Spoilage of fruit juices and juice-containing acidic beverages is expected to increase, as varieties of juice are traded increasingly and internationally. It is therefore emphasized that the juice industry should take appropriate action for controlling these new bacteria, although they are not classified as hazardous but do jeopardize the commercial value of products by their taint production.

The Food Safety Working Group of International Life Sciences Institute Japan (ILSI Japan) has organized a research committee, together with the Japanese soft drink industry, to collect scientific information on the diversity and characteristics of *Alicyclobacillus*. They have been active in the development of effective detection methods and the investigation into their physiological features such as growth temperature and heat resistance in juice products.

The first edition of this book was published in Japanese in 2004. It provided us with a full account of the information currently available on *Alicyclobacillus*. Now I hope that with the English edition, we can share this scientific and technological information and in return encourage international participation in assuring the safety and quality of soft drinks.

June 2006

Mitsukuni Mori, Ph.D.
Former Managing Director, Japan Canners Association

Preface to the Japanese Edition

We are pleased to publish this book, "Thermophilic Acidophilic Bacilli *Alicyclobacillus*," edited by the Microbiology Subcommittee of the Food Safety Research Committee, International Life Sciences Institute Japan (ILSI Japan).

Thermophilic acidophilic bacilli are thermophilic spore formers that can grow at very low pH and at high temperatures such as 40 °C. These bacteria are not inactivated under the conditions in which fruit juice is heat-pasteurized. In addition, they do not produce gas or cause any subsequent change in the appearance of the juice container. Therefore contamination is discovered only by the consumer upon opening and consuming the product. In this respect, thermophilic acidophilic bacilli are difficult bacteria to understand. Spoilage caused by these bacteria occurs not only in Japan but all over the world. They are troublesome bacteria not only for consumers but also for soft drink producers from an ingredient-sourcing and sanitation point of view.

The Microbiology Subcommittee of the Food Safety Research Committee, ILSI Japan, recognized the importance and urgency of this problem, and in the summer of 2003 hosted an international symposium on thermophilic acidophilic bacilli in Tokyo. ILSI Japan consulted with the ILSI headquarters and invited experts from ILSI Europe and ILSI Latin America, who faced problems similar to those in Japan. The unexpectedly large number of participants at this symposium opened our eyes to the seriousness of this problem and led us to publish this book, "Thermophilic Acidophilic Bacilli, *Alicyclobacillus*," which brings together the new insights gained from the symposium with research published thus far.

"Thermophilic Acidophilic Bacilli *Alicyclobacillus*," consisting of the following sections, will be instrumental in understanding the overall picture of this species of bacteria:

(1) Historical background of *Alicyclobacillus*
(2) Characteristics of *Alicyclobacillus*
(3) Methods of detecting *Alicyclobacillus*
(4) Methods of differentiating and identifying *Alicyclobacillus*
(5) Proliferation of *Alicyclobacillus*
(6) Cause of *Alicyclobacillus* contamination and preventive measures
(7) Government offices, associations, and NPOs concerned with *Alicyclobacillus*
(8) Research institutes and researchers of *Alicyclobacillus*

ILSI is a science-based, international NGO, a unique organization working closely with the WHO, and maintains a network within Japan as well as the rest of the world. I am glad to know that this type of publication is made possible by the voluntary efforts of an organization such as ILSI. Last but not least, the enthusiasm and hard work of the Microbiology Subcommittee of the Food Safety Research Committee, ILSI Japan, has made it possible for me to recommend with confidence this valuable book.

October 2004

Shuichi Kimura
Chairman of ILSI Japan

Preface

Since the early 1980s, when spoilage of fruit juices by acid-dependent, thermotolerant spore-forming bacteria was first recognized, members of the genus *Alicyclobacillus* have emerged as food spoilage organisms of major significance to the fruit juice industry. Spoilage is generally manifested as the formation of off-flavors and odors from compounds such as guaiacol and halophenols. The economic impact of such incidents can be very high. To date no human health risks are known to be associated with the consumption of juices and other food products containing *Alicyclobacillus*.

Alicyclobacillus can be difficult to control in fruit juice products as their spores survive juice pasteurization temperatures and may subsequently germinate and grow after processing if conditions are suitable. The spoilage strains of *Alicyclobacillus* grow from around pH 2.5 to 6.0 and at temperatures above 20 °C. Their spores survive for a long period of time in fruit concentrates and similar environments; however, more dilute environments are required for growth. *A. acidoterrestris* is the most commonly occurring species that taints juice and similar products, but other species may also produce taints, particularly in products with a low juice content that are fortified with minerals.

The industry is faced with concern and debate over the significance of *Alicyclobacillus* spp. in raw materials, the behavior of the different species in their products, the necessity for adequate control measures, and the various analytical methods that may detect different species of *Alicyclobacillus* or even the broader group of thermo-acidophilic bacteria.

An inventory by the International Federation of Fruit Juice Producers (IFU) revealed a wide diversity of detection methods in use across the industry and research institutes. Therefore the IFU Working Group on Microbiology took the initiative to develop an internationally acceptable method for the detection of taint-producing *Alicyclobacillus* so that realistic microbiological criteria and specifications can be developed. This has resulted in IFU-Method No.12, for the detection of taint-producing *Alicyclobacillus* in fruit juices, with a first revision of the method appearing in September 2004. Further development of *Alicyclobacillus* detection methods is foreseen as data from ongoing research on the characterization of *Alicyclobacillus* and their behavior in varying substrates become available. International cooperation to harmonize such methods is strongly recommended.

Bob Hartog, M.Sc.
Microbe Control Support
Formerly Chair of Working Group Microbiology
International Federation of Fruit Juice Producers (IFU)

Preface

It is a great privilege to be asked to write a preface for this important publication on *Alicyclobacillus*. The authors of this work have been leaders in the development of detection methodology, identification, and characterization of these spoilage organisms, but until now they have not had the opportunity to publish their body of work in a language other than Japanese. This publication will afford a much larger audience the opportunity to learn and benefit from the research work they have produced.

The microorganisms comprising the genus *Alicyclobacillus* are a very interesting group because they are uniquely adapted to survive and grow in what would normally be considered very harsh environments. The unique properties of these organisms have created the potential for an unusual spoilage problem in certain types of beverage products. These beverages are typically shelf-stable, high-acid, non-carbonated products most commonly packaged in PET containers. The environment that exists in these beverages is generally unwelcoming to most microorganisms, but some *Alicyclobacillus* species find these beverages actually provide a nurturing environment. It is for this reason that these organisms have become a source of concern for the beverage industry and for the raw materials suppliers of that industry.

As a result of this concern, in April 2005, an *Alicyclobacillus* (ACB) subcommittee was formed within the Quality Technical Committee of the International Society of Beverage Technology (ISBT). The subcommittee is composed of three working groups focused on research, methodology, and control. The research working group is reviewing published literature to assess the state-of-the-art, provides on-going updates to the methodology and control groups, and recommends consultants/subject matter experts. The methodology group is reviewing and assessing the development of conventional and rapid methodologies, has created a reference strain collection (of naturally occurring strains), and is working on the publication of standardized test methods. The control group is investigating ingredient/product process flows and sanitizer efficacy with the aim of establishing publishing industry standards/guidelines and best practices.

It is our hope that the publication of this book in English combined with the research efforts of other groups such as ISBT, the IFU, and ILSI Japan, will result in resolving the problems these organisms have created for the beverage industry, its suppliers, and consumers.

Joe Shebuski, Ph.D.
Cargill Global Food Safety& Regulatory Affairs
Cargill, Incorporated

Preface

Some 10 years after the first descriptions of *Alicyclobacillus* in fruit juices by Cerny in 1984, a sample of apparently "normal" orange juice reached our laboratory. The odor alteration was imperceptible, but a faint "strange" smell was detected by chance by one of my colleagues, the first one to open the bottle to smell the content. My client, a major exporter of orange juice, had explained to me the background story of the sample. He had told me that some PDA plates could show a limited growth of Gram-positive bacteria, namely, in a very low number and after a rather long period of incubation, while no growth was detected in ordinary QC methods used for heterotrophic organisms.

As we always do when there is a deviation from the norm, we began a survey of all known acidophilic microorganisms that could have some connection with fruits. At the same time, we tried all kinds of media with acidic pH, or dropped the neutral pH of media with citric acid in an attempt to improve detection. At that time, my suspect was lactobacilli. However, the surface plating resulted in a more successful growth than pour plating, and spores were observed frequently. It was clear that MRS-Lactobacilli media were not their preferred recipe. Successful platings started to be obtained with Orange Serum medium and with acidified PDA at 37 °C, but still the results were not reproducible between counts on different days. Based mostly on the Ph.D. thesis of Deinhard (1987), the study of *Alicyclobacillus* was successfully conducted. I must emphasize that without the precise and detailed information provided by the citrus industries, the early work on *Alicyclobacillus* in orange juices could not have been achieved. They were our reliable partners, collecting samples and evaluating the entire production chain. I am convinced that this first collaborative work, carried out in Brazil about 10 years ago, has developed into a series of worldwide work, which resulted in the present book. I am sure all information contained in this book will be highly welcomed and will be an essential "partner" for those who work in the field of QC and microbiology for acidic juices.

Silvia Yuko Eguchi, Ph.D.
Director of Microbiology (Allergisa)
Formerly R&D Manager, Fundacao Andre Tosello

Preface

With the purpose of providing a source of information on *Alicyclobacillus* that was as complete as possible, current information was compiled and "Thermophilic Acidophilic Bacilli *Alicyclobacillus*" was finally published in Japanese by the publisher Kenpakusha in December 2004. The book enjoyed such great popularity in Japan that, with the hope that it would be of use to researchers in other countries, we embarked on this project to translate it into English, along with the collaboration of several translators and reviewers. It was a long road to completing this book, but finally after 2 years of hard work this English version can be published. We must apologize to the many people who have waited so long for this book. In the present version, we added some new information that was not in the Japanese version.

We sincerely hope this book will be of help to readers in their research and business.

Microbiology Committee, Food Safety Research Division
International Life Sciences Institute of Japan

Contents

Contributors

Joe Shebuski, Ph.D.
Cargill Global Food Safety & Regulatory Affairs
Cargill, Incorporated

Sílvia Y. Eguchi, Ph.D.
Director of Microbiology
Allergisa Pesquisa Dermato-Cosmetica Ltda
Formerly R&D Manager, Colecao de Culturas Tropical
André Tosello Foundation, Brazil

Bob Hartog, M.Sc.
Microbe Control Support
Formerly Chair of Working Group Microbiology
International Federation of Fruit Juice Producers (IFU)

Hisato Ikemoto, Ph.D.
Beverage & Food Division
Suntory Limited

Andrea Kay Suzuki
Hiroyuki Ogawa, Ph.D.
Food Research Laboratories
Mitsui Norin Co., Ltd.

International Life Science Institute of Japan (ILSI Japan)
Kojimachi R·K Building., 2-6-7 Kojimachi, Chiyoda-ku, Tokyo 102-0083, Japan
TEL: +81-3-5215-3535
FAX: +81-3-5215-3537
E-mail: webmaster@ilsijapan.org
Web site: http://japan.ilsi.org/

Authors

Members of Microbiology Division, Food Safety Research Department, ILSI Japan

Keiichi Goto, Ph.D.
Food Research Laboratories
Mitsui Norin Co.,Ltd.

Tetsuya Sawaki
Safety Science Institute
Suntory Limited

Kanjiro Takahashi
Syozo Tanada
Research & Development Division
Nichirei Foods Inc.

Takashi Tanaka
Food Technology Research Institute
Meiji Dairies Corp.

Hajime Tokuda
Safety and Microbial Control Research Center
Kao Corp.

Rie Yamamoto
Teiichi Suzuki
Microbial Science
Coca·Cola Tokyo Research & Development Company Limited

This book is based on the Japanese original edited by the Microbiology Division, Food Safety Research Department, ILSI Japan (Eds.), "Kounetsusei Kousanseikin *Alicyclobacillus*", Kenpakusha Tokyo, 2004 (ISBN 4-7679-9105-X C3045).

Acknowledgments

With the publication of "Thermophilic Acidophilic Bacilli *Alicyclobacillus,*" we wish to express our gratitude to members of the Working Group for Standardized Test Methods for Thermo-Acidophilic Bacilli [Research Committee of the Japan Fruit Juice Association, Asahi Soft Drinks Co., Ltd., Ehime Beverage Inc., Kagome Co., Ltd., Calpis Co., Ltd., Kirin-Tropicana Inc., Kirin Beverage Corp., Suntory Ltd., Coca-Cola (Japan) Co., Ltd., Coca-Cola Tokyo Research & Development Co., Ltd., Nippon Del Monte Corp., and Mitsui Norin Co., Ltd.] of the Japan Fruit Juice Association for permission to insert the abstract of "Standardized Test Methods for Thermo-Acidophilic Bacilli" in this book.

We thank Mr. Bob Hartog (formerly Chair of Working Group, Microbiology, International Federation of Fruit Juice Producers) for his consent to insert the abstract of "IFU-Method No. 12, for the detection of taint-producing *Alicyclobacillus* in fruit juices." Moreover, we deeply thank Mr. Antonio Carlos Gonçalves (Chair of Technical Committee, ABECitrus) for providing much of the valuable data in this book.

We are grateful to Mr. Mikio Furuno, Ms. Momoyo Ishikawa, and Mr Alcides Hideo Tanaka (Nippon Juice Terminal K.K.) for their helpful advice concerning international information. We also wish to convey our thanks to Ecolab K.K., Cuno K.K., Kyokuto Pharmaceutical Industrial Co., Ltd., Nihon Pall Ltd., Nihon Tetra Pak K.K., Mitsubishi-Kagaku Foods Corp., Merck Ltd., and Riken Vitamin Co., Ltd. for providing us with valuable data.

For the publication of this English version of "Thermophilic Acidophilic Bacilli *Alicyclobacillus*" we are extremely indebted to Dr. Sílvia Yuko Eguchi (Allergisa Pesquisa Dermato-Cosmetica Ltda), Dr. Joe Shebuski (Cargill, Inc.) and Ms. Andrea Kay Suzuki (Mitsui Norin Co., Ltd.). We would like to thank the staff of Springer Japan for their help in bringing this book to fruition, and we thank Ms. Tomoko Takanobu and the staff of Kenpakusha for their efforts in publishing the Japanese version. We also thank Sciencedit for reviewing this manuscript.

Finally, we are greatly indebted to Mr. Yuichi Otsuka (Key Coffee Inc.), Mr. Motohiro Niwa (formerly of Kirin Beverage Corp.), Ms. Rieko Fujita (Mitsui Norin Co., Ltd.), and the many contributors to this book.

Acknowledgements

Chapter 1: Introduction

Issues related to spore-forming acidophilic bacteria, *Alicyclobacillus*, are among the most serious problems facing the beverage industry in recent years. This book was written for the purpose of providing a compilation of currently available information about this concern, and it is hoped it will serve as a reference to inquiries on this issue from those involved in beverage production as well as the suppliers of major ingredients such as juice concentrates, liquid sugars and other additives.

Problems with *Alicyclobacillus* have been evident in cases when there was spoilage of juice and juice-containing beverages even after pasteurization at temperatures close to 100°C. The spoilage microorganisms appeared to be spore-forming (heat-resistant) acidophilic rod-shaped bacteria, and they are now named *Alicyclobacillus*. They survive and grow in acidic beverages, and are responsible for the production of guaiacol, a medicine-like phenolic odor. *Alicyclobacillus acidoterrestris*, the most frequent odor producing species, are the main target microorganisms of this quality concern.

While many bacteria prefer to grow under neutral pH conditions, some can survive and grow even in extraordinary habitats under extremely acidic conditions such as in hot springs containing hydrogen sulfide. It is surprising to know, then, that *Alicyclobacillus*, with optimal growth pH at 3.0-5.0, are present quite close to our daily life and are commonly found in soil or fruit samples.

Before getting to the detailed descriptions on *Alicyclobacillus* in the coming chapters, some general information about other acidophilic bacteria, is needed. These organisms are generally defined as those capable of growing at pH conditions lower than 3.0. This is briefly reviewed in Table 1-1. This information is provided by Itoh et al. [1]. and was updated to include *Hydrogenobaculum acidophilum*, *Leptospirillum ferrooxidans*, *Acidiphilum acidophilum*, *Acidiphilum cryptum*, *Acidithiobacillus ferrooxidans*, *Acidithiobacillus caldus*, *Acidithiobacillus albertensis*, *Acidithiobacillus thiooxidans*, *Alicyclobacillus acidocaldarius*, *Alicyclobacillus acidoterrestris*, *Alicyclobacillus acidiphilus*, *Alicyclobacillus herbarius* and *Sulfobacillus thermosulfidooxidans*.

Table 1-1. Characteristics of acidophilic bacteria [2-10)]

Species	Growth pH range	Optimum growth pH	Growth temperature range (°C)	Optimum growth temperature (°C)	Oxgen requirement	Nutrients requirement	Spore	Gram stain	Cell shape	Flagella	G+C contents (%)	Habitats
Hydrogenobacter acidophilus	2.0-6.0	3.0-4.0	40≦ (75°C: no growth)	65	aerobic	autotrophic	-	negative	rod	+	35	hot spring, salfataric field
Leptospirillum ferrooxidans	1.1≦	1.3-2.0	<10-45	30-37	strictly aerobic	autotrophic	-	negative	spiral	+	51-56	acidic mineral environments (<pH3)
Acidiphilum acidophilum	1.5-5.5	2.5-3.0	>25-37	27-30	aerobic	autotrophic /heterotrophic	-	negative	rod	+	63-64	mineral deposit, acidic mineral environments
Acidiphilum cryptum	1.9-5.9		20≦ (47°C: no growth)	35-41	aerobic	heterotorophic	-	negative	rod	+	68-70	mineral deposit, acidic mineral environments
Acidithiobacillus ferrooxidans	1.3-4.5	2.85	10-37	32	strongly aerobic	autotrophic /heterotrophic	-	negative	rod	+	58-59	copper sulfide ores, acidic soil
Acidithiobacillus caldus	1.0-3.5	2.0-2.5	32-52	45	strictly aerobic	autorophic /mixotorophic	-	negative	rod	+	63.1-63.9	abandoned coal mine
Acidithiobacillus albertensis	2.0-4.5	3.5-4.0		28-30	aerobic	autotrophic	-	negative	rod	+	61.5	copper sulfide ores, acidic soil
Acidithiobacillus thiooxidans	0.5-6.0	2.0-2.8	10-37	28-30	strongly aerobic	autotrophic	-	negative	rod	+	50-52	soil, sulfuric spring, acidic mineral deposit
Alicyclobacillus acidocaldarius	2.5-6.0	4.0-4.5	35-70	60	aerobic	heterotorophic	+	variable	rod	+	62	soil, acidic hot spring, fruits juice
Alicyclobacillus acidoterrstris	3.0-6.0	3.5-4.0	20-55	40-50	aerobic	heterotorophic	+	variable	rod	+	51.1-52.7	soil, fruits juice, pectin
Alicyclobacillus acidiphilus	2.5-5.5	3.0	20-55	50	aerobic	heterotorophic	+	variable	rod	+	55.6	material for bevearage
Alicyclobacillus herbarius	3.5-6.0	4.5-5.0	35-65	55-60	aerobic	heterotorophic	+	variable	rod	+	57	herb
Sulfobacillus thermosulfidooxidans	1.1-5.0	1.9-2.4	28-60	50	strongly aerobic	autotrophic /heterotrophic	+	positive	rod/oval	-	53.6-53.9	spontaneously heated ore, Cu-Zn-pyrite deposit

Table 1-2. Characteristics of acido-tolerant bacteria [11-13]

Species	Growth pH	Optimum growth pH	Growth Temperature range (°C)	Optimum Growth Tenperature (°C)	Oxigen requirement	Nutrients requirement	Spore	Gram Stain	Cell shape	Flagella	G+C contents (%)	Habitats
Bacillus coagulans	4.0≦	≦6.0	30-55	37-55	aerobic /anaerobic	heterotrophic	+	positive	rod	+	44.5	acidic foods
Acetobacter aceti	3.2-7.2	5.2-6.4	≦42 (9°C: no growth)	30	strictly aerobic	heterotrophic	-	negative /variable	rod	+	55.9-59.5	brewery, soil
Guluconobacter oxydans	3.6≦	5.5-6.0		25-30	strictly aerobic	heterotrophic	-	negative /variable	rod	+/-	56-64	acetic acid production
Lactobacillus acidophilus	4.0-6.8	5.8-6.6	≦48 (22°C: no growth)	37	microaerobic	heterotrophic	-	positive	rod/oval	-	36	lactic acid production

Table 1-2 summarizes the information on *Bacillus coagulans*, *Acetobacter aceti*, *Gluconobacter oxydans* and *Lactobacillus acidophilus*. They are not acidophiles but known as highly acido-tolerant species. The *Alicyclobacillus* species are, as indicated in Tables 1-1 and 1-2, aerobic, acidophilic and thermophilic bacteria. They are distinguished from the others by their ability to form endospores as is seen with *Bacillus coagulans* (*Alicyclobacillus* species were previously included in the genus *Bacillus*.).

Tetsuya Sawaki

References

1. Ito Y, Ito T (2002) Biseibutu riyou no daitenkai. In NTS., pp 113-121 (in Japanese).
2. Stör R, Waberski A, Vöker H, Tindall BJ, Thomm M (2001) *Hydrogenothermus marinus* gen. nov., sp. nov., a novel thermophilic hydrogen-oxidizing bacterium, recognition of *Calderobacterium hydrogenophilum* as a member of the genus *Hydrogenobacter* and proposal of the reclassification of *Hydrogenobacter acidophilus* as *Hydrogenobaculum acidophilum* gen. nov., comb. nov., in the phylum '*Hydrogenobacter/Aquifex*'. Int J Syst Evol Microbiol 51:1853-1862.
3. Shima S, Suzuki K (1993) *Hydrogenobacter acidophilus* sp. nov., a thermoacidophilic, aerobic, hydrogen-oxidizing bacterium requiring elemental sulfur for growth. Int J Syst Bacteriol 43:703-708.
4. Harrison AP (1981) *Acidiphilum cryptum* gen. nov., heterotrophic bacterium from acidic mineral environments. Int J Syst Bacteriol 31:327-332.
5. Harrison AP (1983) Genomic and physiological comparisons between Heterotrophic Thiobacilli and *Acidiphilum cryptum*, *Thiobacillus versutus* sp. nov., and *Thiobacillus acidophilus* nom. rev. Int J Syst Bacteriol 33:211-217.
6. Kinsel NA (1960) New sulfur oxidizing iron bacterium: *Ferrobacillus sulfooxidans* sp. nov. J Bacteriol 80:628-632.
7. Kelly DP, Wood AP (2000) Reclassification of some species of *Thiobacillus* to the newly designated genera *Acidithiobacillus* gen. nov., *Halothiobacillus* gen. nov. and *Thermothiobacillus* gen. nov. Int J Syst Evol Microbiol 50:511-516.
8. Hallberg KB, Lindström EB (1994) Characterization of *Thiobacillus caldus* sp. nov., a moderately thermophilic acidophile. Microbiol 140:3451-3456.
9. Bryant RD, McGroarty KM, Costerton JW, Laishley EJ (1983) Isolation and characterization of a new acidophilic *Thiobacillus* species (*T. albertis*). Can J Microbiol 29:1159-1170.
10. Golovacheva RS, Karavaiko GI (1978) A new genus of thermophilic spore-forming bacteria, *Sulfobacillus*. Microbiol 47:658-664.

11. Ohmori S, Masai H, Arima K, Beppu T (1980) Isolation and identification of acetic acid bacteria for submerged acetic acid fermentation at high temperature. Agric Biol Chem 44:2901-2906.
12. Todar K (2001) Nutrition and growth of bacteria. University of Wisconsin-Madison Department of Bacteriology, Wisconsin, pp 11-12.
13. Aiuchi M, Hirota T, et al. (1980) Yoguruto: Sono kisokagaku oyobi gijutsu. Seizo to Chouseihou. In Jituyotosho, p 80 (in Japanese).

Chapter 2: Historical background related to *Alicyclobacillus*

The main historical facts related to *Alicyclobacillus* are as follows:

1967 : Uchino and Doi of Nagoya University (Aichi Pref., Japan) reported the isolation of a spore-forming bacteria from a hot spring site near Lake Tazawa (Iwate Pref., Japan), that grew under acidic and geothermal conditions, and has very similar characteristics to *Bacillus coagulans* [1)].

1971 : Darland and Brock isolated bacteria from variety of thermal acidic environments, both aqueous and terrestrial at Yellowstone National Park and Hawaiian Volcano National Park, USA, with very similar characteristics to the bacteria reported by Uchino and Doi [2)]. Based on the taxonomical properties, they named it *Bacillus acidocaldarius*; a new species of the genus *Bacillus*. They also discovered this microorganism possessed unusual ω-cyclohexyl fatty acids as its main cellular fatty acid component.

1981 : Hippchen and co-workers isolated a thermo-acidophilic spore-forming bacteria from soil and concluded it had widespread distribution. Later on, this bacterium was studied in detail by Deinhard and co-workers, and in 1987 was proposed as a new species, *Bacillus acidoterrestris* [3, 4)].

1982 : Contamination of pasteurized apple juice occurred in Germany [5)] on a large scale. This was the first reported incidence of spoilage by *Alicyclobacillus*. The cause was attributed to an organism related to *Bacillus acidocaldarius*. However, later studies showed that the cause was *Alicyclobacillus acidoterrestris*. Also, around 1980, these authors detected a foul odor in an experimental sample of a combined plum and apple juice. In retrospect, we can now probably assume this was caused by *Alicyclobacillus acidoterrestris*.

1989 : The first report in Japan of an incident of deterioration in an acidic juice product due to a similar bacterium was made by Suzuki [6)].

1990 & 1995 : In Australia, an unpleasant odor detected in a 40 % apple juice sample (TetraPack, 250 mL) was attributed to the presence of 2,6-dibromophenol and 2,6-dichlorophenol. It was reported this bad odor was not caused by any additives or preservatives added to the juice, but by microbial contamination [7].

1991 : Following reports by Niwa and co-workers of a series of contamination incidents, the problem started drawing attention in Japan [8].

1994 : Splittstoesser and co-workers reported the strain isolated from pasteurized apple juice in 1990 was a thermo-acidophilic bacterium. This was the first mention of an isolation of this organism from contaminated fruit juice in the USA [9].

1995 : McIntyre and co-workers isolated thermo-acidophilic bacteria from a variety of fruit juices in the USA [10].

1996 : In Japan, Yamazaki and co-workers isolated and identified *Alicyclobacillus acidoterrestris* from a spoiled acidic drink [11].

1997 : In Australia, a high concentration of guaiacol was detected in a 25 % tropical fruit drink, from which an *Alicyclobacillus* related strain was isolated. The contamination source was reported to be an apple juice [7].

From this time on, research into the isolation, characterization, methodology for detection, rapid methods for differentiation, process control technology and development of control techniques, has actively been pursued in Japan, the United States, the European Community, Brazil and Australia, with a number of reports having since been published.

Tetsuya Sawaki

References

1. Uchino F, Doi S (1967) Acido-thermophilic bacteria from thermal waters. Agric Biol Chem 31:817-822.
2. Darland G, Brock TD (1971) *Bacillus acidocaldarius* sp. nov., an acidophilic thermophilic spore-forming bacterium. J Gen Microbiol 67:9-15.
3. Hippchen B, Röll A, Poralla K (1981) Occurrence in soil of thermo-acidophilic bacilli possesing ω-cyclohexane fatty acids and hopanoids. Arch Microbiol 129:53-55.

4. Deinhard G, Blanz P, Poralla K, Altan E (1987) *Bacillus acidoterrestris* sp. nov., a new thermotolerant acidophile isolated from different soils. System Appl Microbiol 10:47-53.
5. Cerny G, Hennlich W, Poralla K (1984) Fruchtsaftverderb durch Bacillen: Isolierung und Charakteriesierung des Verderbserregers. Z Lebens Unters Forsch 179:224-227 (in German).
6. Suzuki T (1989) Sansei inryo kanzume no kousansei kin. Shokuhin to yoki 30:503-506 (in Japanese).
7. Goldberg D (2002) Experiences with *Alicyclobacillus* (ACB). Berri Limited, Australia/New Zealand.
8. Niwa M, Hirota Y, Shirosu Y (1991) Yunyu genryo kaju yori kenshutsu sareta shinki tainetusei kousansei kin. Kaju-Kyokai Hou 9:31-42 (in Japanese).
9. Splittstoesser DF, Churey JJ, Lee CY (1994) Growth characteristics of aciduric sporeforming bacilli isolated from fruit juices. J Food Protection 57:1080-1083.
10. McIntyre S, Ikawa JY, Parkinson N, Haglund J, Lee J (1995) Characteristics of an acidophilic *Bacillus* strain isolated from shelf-stable juices. J Food Protection 58:319-321.
11. Yamazaki K, Tezuka H, Shinano H (1996) Isolation and identification of *Alicyclobacillus acidoterrestris* from acid beverages. Biosci Biotech Biochem 60:543-545.

Chapter 3: Characteristics of *Alicyclobacillus*

3-1 Abbreviations used for *Alicyclobacillus*

TAB, the abbreviation for "Thermo Acidophillic Bacilli (or Bacteria)" is the most widely utilized way to designate the bacteria belonging to genus *Alicyclobacillus*, and while it is not an exact label for *Alicyclobacillus*, it has come to be accepted as synonym. The designation adopted in Japanese is equivalent to "Thermo-tolerant Acidophillic Bacilli (or Bacteria)", which is also abbreviated as TAB. Some moderately thermophilic acidophilic bacteria such as *Bacillus fumarioli* are included in the abbreviation, TAB, in a broader sense. Other abbreviations commonly used internationally are AAT for *Alicyclobacillus acidoterrestris* and BAT, the former name for *Bacillus acidoterrestris*. This is still used for some designations such as culture media. ATSB is sometimes seen as an abbreviation for "Acido-thermophilic Spore-forming Bacteria", but is rarely used in Japan. Recently, ACB has been used as the abbreviation of *Alicyclobacillus* worldwide. These abbreviations are routinely used in language related to quality control and assurance. When a species and/or strain name is clearly specified or used for technical purposes, the scientific name of the microorganism is adopted. Although somewhat confusing, these are the current designations in use. The abbreviations used in this book follow those of the original papers.

3-2 Classification

In 1967, Uchino and Doi isolated aerobic spore-forming bacteria from a hot spring in Japan using acidic and high temperature conditions. They reported the organisms as new species, closely related to *Bacillus coagulans* [1)]. While a clear confirmation of the taxonomic status of the isolates is not possible now because they are not preserved, the majority of their microscopic, physiological and biochemical characteristics resemble those of the bacteria now known as *Alicyclobacillus* [2)], indicating that this was the first report of an *Alicyclobacillus* isolation. A few years later, Darland and

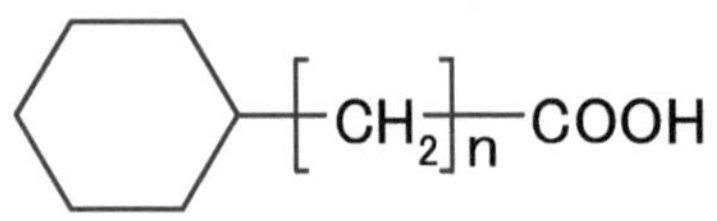

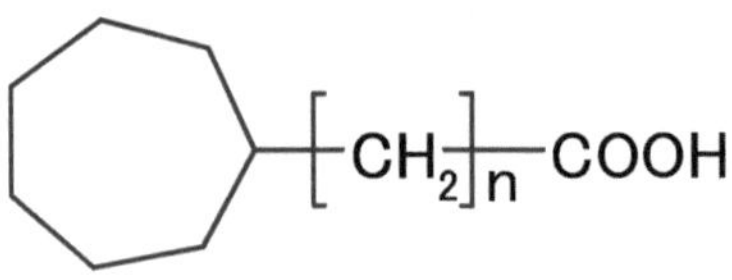

ω-cyclohexyl fatty acids ω-cycloheptyl fatty acids

Fig. 3-1. ω-alicyclic fatty acids

Brock isolated bacteria with characteristics very close to those of the strains, which were isolated by Uchino and Doi, from soil and water samples collected from Yellowstone National Park and the Hawaiian Volcano National Park, USA. Based on the results of taxonomical research in 1987, they were classified as a new species, *Bacillus acidocaldarius* [3]. Along with their thermophilic and acidophilic characteristics, these bacteria possessed a cyclic alkyl terminal (alicyclic) radical in their cellular fatty acid structure. These very distinctive ω-cyclohexyl fatty acids (Fig. 3-1) are the main component of the essential fatty acids of these organisms [4]. In 1987, microorganisms with ω-cyclohexyl fatty acids were isolated from soil in Europe and spoiled apple juice, but had some differing characteristics from *Bacillus acidocaldarius*, and were named *Bacillus acidoterrestris* [5]. Other isolates, also from soil, possessed ω-cycloheptyl fatty acids (Fig. 3-1), and were named *Bacillus cycloheptanicus* [6]. In 1992, Wisotzkey and co-workers reported that *B. acidocaldarius*, *B. acidoterrestris* and *B. cycloheptanicus* should be reclassified from *Bacillus* to a new genus, *Alicyclobacillus*, according to these three species possess the unique ω-alicyclic fatty acids and form one cluster based on the phylogenetic analysis of 16S rRNA gene (16S rDNA) sequence [2]. For some time after, *Alicyclobacillus* was represented by these three species. Recently though, *Alicyclobacillus* species have been identified as the cause of off odors associated with the spoilage of acidic beverages, and reported not only in Japan [7-16], but also in other parts of the world. Before this, incidents of spoilage linked to *Alicyclobacillus* were few and only sporadically reported [17-20]. There was a great deal of speculation which attempted to explain these incidents including global warming, acid rain, large scale production/factories and other factors, and as a result, these spoilage incidents fueled research work on the isolation and identification of *Alicyclobacillus*. A number of *Alicyclobacillus* isolates were found in soil, hot springs, beverages and raw materials, and the following species were added as members of *Alicyclobacillus*: "*A. mali*" [21], *A. acidocaldarius* subsp. *rittmannii* [22], *A. hesperidum* [23], *Alicyclobacillus* genomic species 1 [23], *A. acidiphilus*

[16], *A. herbarius* [15], *Alicyclobacillus* genomic species 2 [24], *A. sendaiensis* [25], *A. pomorum* [26] and *A. vulcanalis* [87]. Recently, *Sulfobacillus thermosulfidooxidans* subsp. *thermotolerans* and *Sulfobacillus disulfidooxidans* were reclassified as *A. tolerans* [86] and *A. disulfidooxidans* [86], respectively. Now *Alicyclobacillus* is comprised of twelve species (including one subspecies), two genomic species and one proposed species ("*A. mali* "). It is also possible that *Alicyclobacillus* genomic species 1 belongs to "*A. mali* ", and *Alicyclobacillus* genomic species 2 belongs to *A. sendaiensis* (Table 3-1). Taxonomical research on *Alicyclobacillus* has greatly improved the organization of this group compared to the past. However, factors such as phylogenetic reassociation and further addition of a large number of new species (the position of the novel species in the phylogenetic tree is shown in Fig. 3-2), need to be considered from the taxonomical point of view at greater depth.

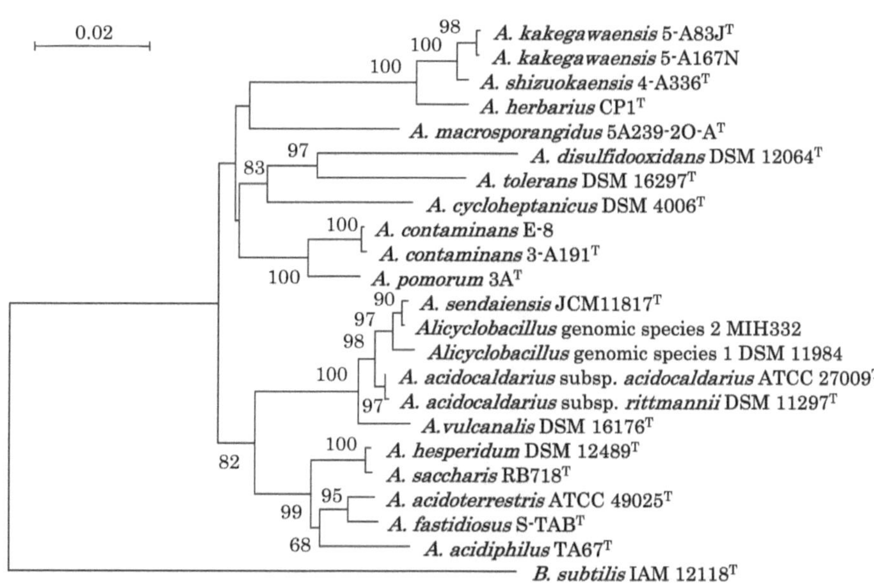

Fig. 3-2. Phylogenetic tree (constructed using the maximum parsimony method) of *Alicyclobacillus* species and related species based on 16S rDNA

Bacillus subtilis IAM 12118^T was served as outgroup. The final dataset included 1468 unambiguously aligned sites. Numbers represent percentages from 1000 replicate bootstrap samplings (frequencies of less than 70 % are not shown). Bar, 0.02 substitutions per nucleotide position.

Table 3-1. *Alicyclobacillus* species and the sources

Species	Description	Source
A. acidocaldarius subsp. *acidocaldarius*	Darland & Brock, 1971 [3]	Soil, raw materials of beverage, etc.
A. acidocaldarius subsp. *rittmannii*	Nicolaus et al., 1998 [22]	Soil
Alicyclobacillus genomic species 1	Albuquerque et al., 2000 [23]	Soil, raw materials of beverage, etc.
"*A. mali* "	Kusano et al., 1997 [21]	Apple juice
A. sendaiensis	Tsuruoka et al., 2003 [25]	Compost
Alicyclobacillus genomic species 2	Goto et al., 2002 [24]	Soil
A. vulcanalis	Simbahan et al., 2004 [90]	Soil
A. acidoterrestris	Deinhard et al., 1987 [5]	Soil, raw materials of beverage, etc.
Alicyclobacillus sp. 3 (= *A. fastidiosus*)	Goto et al., in press [89]	Raw material of beverage
A. acidiphilus	Matsubara et al., 2002 [16]	Raw material of beverage
A. hesperidum	Albuquerque et al., 2000 [23]	Soil
Alicyclobacillus sp. 1 (= *A. saccharis*)	Got et al., in press [89]	Raw material of beverage, etc.
Alicyclobacillus sp. 2	Unpublished	Soil
A. cycloheptanicus	Deinhard et al., 1987 [6]	Soil
Alicyclobacillus sp. (= *A. macrosporangidus*)	Got et al., in press [89]	Soil
A. pomorum	Goto et al., 2003 [26]	Fruit juice
Alicyclobacillus sp. 4 (= *A. contaminans*)	Got et al., in press [89]	Soil
Alicyclobacillus sp. 5 (= *A. contaminans*)	Got et al., in press [89]	Raw material of beverage
Alicyclobacillus sp. 6 (= *A. kakegawaensis*)	Got et al., in press [89]	Soil
Alicyclobacillus sp. 7 (= *A. kakegawaensis*)	Got et al., in press [89]	Soil
Alicyclobacillus sp. 8 (= *A. shizuokaensis*)	Got et al., in press [89]	Soil
A. herbarius	Goto et al., 2002 [15]	Herb
Alicyclobacillus sp. 9 (= *A. herbarius*)	Got et al., in press [89]	Soil
A. disulfidooxidans	Karavaiko et al., 2005 [91]	Soil
A. tolerans	Karavaiko et al., 2005 [91]	Soil

3-3 Characteristics of *Alicyclobacillus*

This section will discuss the overall characteristics and the primary differentiation properties of *Alicyclobacillus*. Other detailed descriptions can be found in the original references [2, 3, 5, 6, 15, 16, 22, 23, 24, 25, 26, 86, 87].

3-3-1 Morphology

Vegetative cells of *Alicyclobacillus* are rod shaped with terminal/subterminal endospores that are approximately 0.7-1 μm wide and 3-5 μm long. The swelling of the cells due to endospore formation may or may not occur, depending on the species (a greater number of strains show swelling, Fig. 3-3). Chain formation is rare and motility is weak. The cells stain as Gram-positive in the early stage of the cultivation and become Gram-negative or Gram-variable at the end of the cultivation. This fact was also confirmed by applying the Ryu method [27], where the cells stained Gram-positive in younger cultures and Gram-negative with older cultures. The size of the colonies is dependent on the growth medium, reaching 2-5 mm (sometimes larger), on YSG agar medium at optimum temperature. The colony morphology can differ slightly depending on the strain, but in general round colonies are formed (Fig. 3-3). Some older cultures may appear contaminated, due to their heterogeneous morphology. The colonies range in color from white to beige, becoming slightly darker with age. Their sheen is not quite glossy. Growth beneath the surface of the agar medium is limited to approximately 0.5-1 mm because of low oxygen concentration.

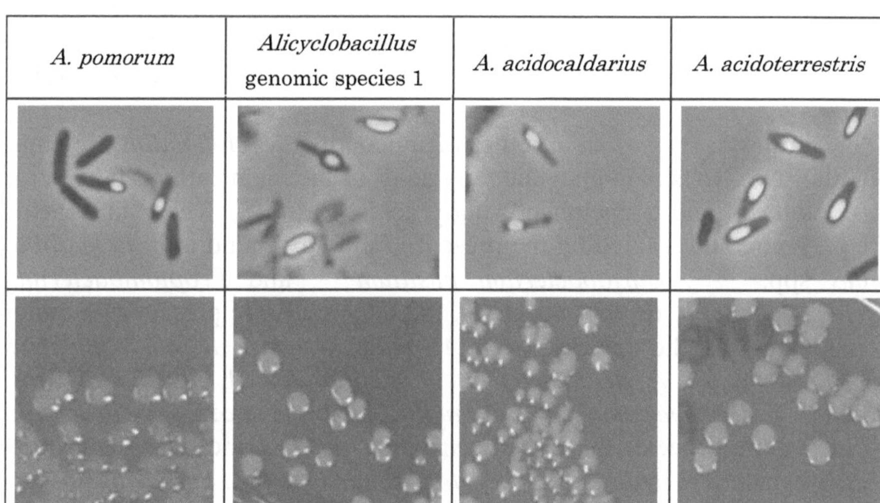

Fig. 3-3. Morphological characteristics of 4 *Alicyclobacillus* species

All strain were grown on YSG agar.

3-3-2 Physiology and biochemical properties

Alicyclobacillus can grow in a temperature range from 20-70°C with an optimum growth temperature: between 40 and 60°C and a pH range 2.0-6.0 with an optimum between 3.5 and 4.5 (Table 3-2). Growth is reduced at temperature and pH values outside the optimum range. Although strict aerobes, these organisms can survive under micro-aerobic conditions (growth is inhibited in the absence of oxygen). They metabolize sugars very well, with acid production (Table 3-3). The sugars that can be metabolized tend to form a group according to the species, although there is a rather large variation within species [92)]. The metabolism of sugars is not accompanied by gas production. Depending on the nutritional state of the cells, the presence of salt, organic acids, polyphenols and alcohols, over a certain concentration can inhibit growth, however some strains are known to exhibit resistance to these inhibitors (See Chapter 8).

3-3-3 Quinone species and the composition of cellular fatty acids

As in *Bacillus*, the major quinone (related to the respiratory chain) of *Alicyclobacillus* is MK-7 (menaquinone-7), although some strains have a quinone with a shorter chain, such as MK-3. The relative amount of these quinones is dependent on the strain and the conditions of cultivation (Table 3-4).

Excluding *A. pomorum*, all other members of *Alicyclobacillus* are composed of two types of ω-alicyclic fatty acids as the main cellular fatty acids (ω-cyclohexyl and ω-cycloheptyl fatty acids; Fig. 3-5). These fatty acids are very rare and, excluding the present group, are found only in *Sulfobacillus* spp. [28-30)], *Curtobacterium pusillum* [31)] and *Propionibacterium cyclohexanicum* [32)] (in these bacteria, the fatty acids are present as a minor component). At the species level, *A. acidocaldarius*, *A. acidocaldarius* subsp. *rittmannii*, *A. acidoterrestris*, *A. acidiphilus*, *A. disulfidooxidans*, "*A. mali* ", *A. hesperidum*, *Alicyclobacillus* genomic species 1, *Alicyclobacillus* genomic species 2, *A. sendaiensis*, *A. tolerans* and *A. vulcanis* possess ω-cyclohexyl fatty acids. *A. cycloheptanicus* and *A. herbarius* possess ω-cycloheptyl fatty acids. *A. pomorum* does not have any of these fatty acids, possessing iso and anteiso branched fatty acids. The presence of ω-alicyclic fatty acids is thought to be the factor that enables these organisms to survive under acidic and thermal conditions [33, 34)], but it is not an indispensable condition for survival.

Table 3-2. Growth temperature and pH range of *Alicyclobacillus* species [25, 26, 86, 87, 89]

Species	Growth Temperature range (°C)	Optimum Growth Temperature (°C)	Growth pH range	Optimum pH
A. acidocaldarius subsp. *acidocaldarius*	35-70	55-60	2.5-6.0	4.5-5.0
A. acidocaldarius subsp. *rittmannii*	45-70	63	2.5-5.0	4.0
Alicyclobacillus genomic species 1	35-70	55-60	2.0-6.0	4.0-4.5
"*A. mali*"	35-65	50	2.0-5.5	4.0-5.0
A. sendaiensis	40-65	55	2.5-6.5	5.5
Alicyclobacillus genomic species 2	35-70	55-60	2.5-6.0	4.0-4.5
A. vulcanalis	35-65	55	2.0-6.0	4.0
A. acidoterrestris	20-55	40-50	3.0-6.0	3.5-4.0
Alicyclobacillus sp. 3 (= *A. fastidiosus*)	20-55	40-45	2.0-5.5	4.0-4.5
A. acidiphilus	20-55	50	2.5-5.5	3.0
A. hesperidum	35-60	50-53	2.5-5.5	3.5-4.0
Alicyclobacillus sp. 1 (= *A. saccharis*)	30-55	45-50	2.0-6.0	4.0-4.5
Alicyclobacillus sp. 2	Unknown	Unknown	Unknown	Unknown
A. cycloheptanicus	30-55	50	3.0-5.5	4.0
Alicyclobacillus sp. (= *A. macrosporangidus*)	35-60	50-55	3.0-6.5	4.0-4.5
A. pomorum	30-60	45-50	2.5-6.5	4.5-5.0
Alicyclobacillus sp. 4 (= *A. contaminans*)	35-60	50-55	3.0-6.0	4.0-4.5
Alicyclobacillus sp. 6 (= *A. kakegawaensis*)	40-60	50-55	3.0-6.5	4.0-4.5
Alicyclobacillus sp. 8 (= *A. shizuokaensis*)	35-60	45-50	3.0-6.5	4.0-4.5
A. herbarius	35-65	55-60	3.5-6.0	4.5-5.0
A. disulfidooxidans	4-40	35	0.5-6.0	1.5-2.5
A. tolerans	≦20-55	37-42	1.5-5.0	2.5-2.7

Table 3-3. Differencial characteristics of known *Alicyclobacillus* species

Characteristic	1	2	3 [23]	4	5	6 [23]	7 [16]	8	9	10 [25]
Oxidase	+	-	-	-	-	-	-	+	-	-
Catalase	+	+	+	+	+	+	+	+	+	-
Gelatin hydrolysis	+	+	+	+	+	+	-	-	-	n.d.
Starch hydrolysis	+	+	+	+	-	+	-	-	-	n.d.
Nitrate reduced to nitrite	-	-	-	+	-	-	-	-	+	+
Growth in 5% NaCl	+	+	+	+	+	+	+	+	+	+
Growth in 5% NaCl	-	-	-	-	+	-	-	+	+	n.d.
Acid production from:										
Glycerol	+	+	+	+	+	+	-	-	+	+
Erythritol	-	-	-	-	+	-	-	-	-	-
D-Arabinose	-	-	-	-	-	-	+	+	+	-
L-Arabinose	-	+	+	+	+	+	+	+	+	+
D-Xylose	-	+	+	+	-	-	+	+	+	+
L-Xylose	-	-	-	-	-	-	-	+	-	-
D-Galactose	-	+	+	+	+	+	+	-	+	+
L-Sorbose	+	-	-	+	-	-	+	+	-	-
Rhamnose	-	+	+	+	+	-	-	+	+	-
Inositol	-	-	-	-	+	-	-	-	-	-
Mannitol	+	+	+	+	+	+	-	+	+	+
Sorbitol	-	-	-	-	+	-	+	+	-	-
Methyl α-D-mannoside	-	-	-	+	-	-	-	-	+	-
Methyl α-D-glucoside	+	+	-	+	-	-	+	-	+	+
Amygdalin	+	-	-	-	-	-	-	-	+	-
Arbutin	-	+	+	+	-	-	+	-	+	+
Aesculin	+	+	-	+	-	-	+	+	+	-
Salicin	+	+	+	+	-	-	+	-	+	+
Cellobiose	-	+	+	+	+	+	+	-	+	+
Maltose	+	+	+	+	+	+	+	-	+	+
Lactose	-	+	+	+	+	+	+	-	+	+
Melibiose	-	+	-	+	w	-	-	-	+	-
Sucrose	+	+	+	+	+	+	+	-	+	+
Trehalose	+	+	+	+	+	+	+	-	+	+
Melezitose	-	-	+	+	-	-	+	-	+	-
D-Raffinose	-	+	+	+	-	-	+	-	+	+
Starch	-	+	-	+	-	w	-	-	-	n.d.
Glycogen	-	+	+	+	-	+	-	-	-	+
Xylitol	-	-	-	+	+	-	+	-	-	-
β-Gentiobiose	-	-	+	+	-	w	+	-	+	-
D-Turanose	+	-	+	+	-	+	+	-	+	+
D-Lyxose	-	-	-	-	-	-	-	+	-	-
D-Tagatose	+	-	+	+	-	-	-	+	-	-
D-Fucose	-	-	-	-	-	-	-	-	+	-
5-Keto-gluconate	+	-	-	-	-	-	-	+	+	-

Strain 1, *A. pomorum* 3A^{T}; 2, *A. acidocaldarius* subsp. *acidocaldarius* ATCC 27009^{T}; 3, *Alicyclobacillus* genomic species 1 DSM 11984; 4, *Alicyclobacillus* genomic species 2 MIH332; 5, *A. acidoterrestris* ATCC 49025^{T}; 6, *A. hesperidum* DSM 12489^{T}; 7, *A. acidiphilus* TA 67^{T}; 8, *A. cycloheptanicus* DSM 4006^{T}, 9, *A. herbarius* IAM 14883^{T}; 10, *A. sendaiensis* JNTAP-1^{T}. +: positive, -: negative, w: weakly positive, n.d.: no data avairable. All strain positive: D-ribose, D-glucose, D-fructose and D-mannose. All strain negative: adonotol, methyl β-xyloside, dulcitol, N-acetyl-glucosamine, L-fucose, L-arabitol, gluconate, inulin and 2-keto-gluconate.

Table 3-4. Menaquinone composition of strains of *Alicyclobacillus acidocaldarius* group [25, 26, 86, 87, 89)]

Strain	MK3 (%)	MK7 (%)
Alicyclobacillus genomic species 1 DSM 11984	5.2	94.8
Alicyclobacillus genomic species 1 DSM 11983	4.5	95.5
Alicyclobacillus genomic species 1 DSM 453	17.7	82.3
Alicyclobacillus genomic species 1 DSM 454	6.5	93.5
Alicyclobacillus genomic species 1 DSM 455	9.2	90.8
Alicyclobacillus genomic species 1 DSM 12490	4.7	95.3
Alicyclobacillus genomic species 1 DSM 6481	4.0	96.0
A. acidocaldarius subsp. *rittmannii* DSM 11297^T	7.7	92.3
A. acidocaldarius subsp. *acidocaldarius* ATCC 27009^T	29.0	71.0
A. acidocaldarius DSM 448	11.9	88.1
A. acidocaldarius DSM 449	12.1	87.9
A. acidocaldarius DSM 451	26.8	73.2
A. acidocaldarius DSM 452	9.1	90.9
A. sendaiensis JNTAP-1^T [25)]	0	100
Alicyclobacillus genomic species 2 MIH 332	2.0	98.0

3-3-4 DNA G+C content and DNA-DNA similarity

The range of the Guanine + Cytosine content (G+C mol%) in the chromosomal DNA is 53-63 mol% [26)]. This value is near 62 % for *A. acidocaldarius* and *Alicyclobacillus* genomic species 1 [15, 24, 35)], and the values of the other species are near 55 % (Table 3-6).

The DNA-DNA relatedness among *Alicyclobacillus* spp. is 10-70 % between species (50-70 % between subspecies and under 50 % for different species) [36)]. The relatedness between *A. acidocaldarius* and *Alicyclobacillus* genomic species 1 is approximately 50-70 %, a difference typically obtained between subspecies and genotypes, indicating the close similarity between these two strains (Table 3-6).

3-3-5 16S rRNA gene (16S rDNA)

The similarity of the 16S rDNA nucleotide sequence within the genus *Alicyclobacillus* is reported to be over 92 % [2)]. Within close species, especially belonging to *A. acidocaldarius* group, the similarity is over 98 % [15, 24, 35)]. The phylogenetic tree obtained from the 16S rDNA sequence analysis is shown in Fig. 3-5. Recently, 16S rDNA sequence is routinely used, not only for *Alicyclobacillus*, but also for classification and identification of various bacteria. However, it is important to keep in mind that for the calculation of similarity or to perform a phylogenetic analysis, a reliable database is needed containing sequences that can be trusted (databases consisting of sequences that are not confirmed should not be used), and the determined nucleotide sequence must of course be generated by high quality experimental data.

Table 3-5. Cellular fatty acid composition of type/reference strains of *Alicyclobacillus* species [89)]

Fatty acid	1	2	3	4	5	6	7	8	9	10	11	12	13	14	15	16	17	18	19	20
$C_{15:0}$	tr											1.2	0.5					0.6		n.d.
$C_{16:0}$	4.9	1.7	3.2	1.9	0.5	7.8	0.5	1.7	2.2	1.3	0.8	8.8	6.1	3.5	6.1	0.9	0.7	2.4	0.8	n.d.
$C_{17:0}$	0.6											1.2	1.6	0.6		0.6				n.d.
$C_{18:0}$	0.9	2.1	1.5	2.1										1.0		0.6		1.3		n.d.
iso-$C_{15:0}$	1.1	1.4	1.5	2.4	2.6	1.5	0.8	0.5	0.6	2.5	tr	tr	0.7	tr	2.1	1.8	10.6	0.5	0.7	n.d.
iso-$C_{16:0}$	1.6	0.8	2.8	1.8	2.3	4.2		0.5		8.3		0.9	1.9	4.7	44.2	16.9	20.3	1.0	11.3	n.d.
iso-$C_{17:0}$	4.5	2.6	5.1	5.1	10.1	8.3	2.4	2.0	1.2	5.3	1.2	0.5	4.0	5.5	16.7	29.3	11.1	2.7	0.8	n.d.
iso-$C_{18:0}$										tr				2.9	4.5	4.7	0.5			n.d.
iso-$C_{19:0}$																0.9				n.d.
anteiso-$C_{15:0}$	1.5	0.5	1.2	0.9	tr	0.7	0.5	1.1	0.5	1.9	tr				1.2	0.5	11.9	tr	1.9	n.d.
anteiso-$C_{17:0}$	10.5	0.7	6.7	2	1.8	8.5	4.6	6.5	3.2	5.4	3.7			2.3	25.2	43.8	44.9	1.3	17.4	n.d.
ω-Cyclohexane $C_{17:0}$	44.1	47.6	52.8	52.4	51.1	46.2	64.4	67.8	68.4	61.1	61.7								42.7	27.5-60.0
ω-Cyclohexane $C_{17:0\ 2OH}$																				8.2-11.3
ω-Cyclohexane $C_{19:0}$	30.2	42.6	25.2	31.4	31.3	22.8	26.9	19.9	23.9	13.9	31.7								7.2	2.1-7.9
ω-Cycloheptane $C_{18:0}$												65.6	65.4	55.3				86.8		
ω-Cycloheptane $C_{18:0\ 2OH}$												15.0	12.4	9.7				2.3		
ω-Cycloheptane $C_{20:0}$												6.4	7.4	14.2				tr		

Strain 1, *A. sendaiensis* JCM 11817^T; 2, *Alicyclobacillus* genomic species 2 MIH332; 3, *Alicyclobacillus* genomic species 1 DSM 11984; 4, *A. acidocaldarius* subsp. *rittmannii* DSM 11297^T; 5, *A. acidocaldarius* subsp. *acidocaldarius* ATCC 27009^T; 6, *A. vulcanalis* DSM 16176^T; 7, *A. saccharis* RB718^T; 8, *A. hesperidum* DSM 12489^T; 9, *A. acidiphilus* TA 67^T; 10, *A. fastidiosus* S-TAB^T; 11, *A. acidoterrestris* ATCC 49025^T; 12, *A. kakegawaensis* 5-A83J^T; 13, *A. shizuokaensis* 4-A336^T; 14, *A. herbarius* IAM 14883^T; 15, *A. macrosporangidus* 5-A239-2O-A^T; 16, *A. contaminans* 3-A191^T; 17, *A. pomorum* 3A^T; 18, *A. cycloheptanicus* DSM 4006^T; 19, *A. disulfidooxidans* DSM 12064^T; 20, *A. tolerans* DSM 16297^T. tr, trace amount (<0.5 %); n.d., no data.

Table 3-6a. DNA base compositions and DNA-DNA hybridization values between isolates and type strains of related *Alicyclobacillus* species [25, 26, 89]

Strain	DNA-DNA similarity (%)									
	1	2	3 [25]	4	5	6	7	8	9	10
Alicyclobacillus genomic species 1 DSM 11984	100	66		16	21	20	15	13	10	8
Alicyclobacillus genomic species 1 DSM 11983	99	57								
Alicyclobacillus genomic species 1 DSM 453	93	65								
Alicyclobacillus genomic species 1 DSM 454	89	61								
Alicyclobacillus genomic species 1 DSM 455	86	57								
Alicyclobacillus genomic species 1 DSM 12490	90	63								
Alicyclobacillus genomic species 1 DSM 6481	79	59								
A. acidocaldarius subsp. *rittmannii* DSM 11297^T	58	81							10	
A. acidocaldarius subsp. *acidocaldarius* ATCC 27009^T	57	100	33	20	16	29	13	10	12	10
A. acidocaldarius DSM 448	61	92								
A. acidocaldarius DSM 449	60	99								
A. acidocaldarius DSM 451	51	95								
A. acidocaldarius DSM 452	52	93								
A. sendaiensis JNTAP-1^T			100							
Alicyclobacillus genomic species 2 MIH 332	44	39								
A. acidoterrestris ATCC 49025^T	12	11	5	100	23	24	15	12	10	11
A. acidoterrestris DSM 3923				86	20	26				
A. acidoterrestris DSM 3924				89	19	26				
A. acidoterrestris DSM 2498				75	27	23				
A. acidiphilus TA 67^T	14	15		20	100	31	22	12	11	9
A. hesperidum DSM 12489^T	23	24	26	30	25	100	13	10	11	12
A. hesperidum DSM 12766				32	21	92				
A. cycloheptanicus DSM 4006^T	12	11		13	21	19	100	12	13	8
A. cycloheptanicus DSM 4005							86			
A. cycloheptanicus DSM 4007							79			
A. herbarius CP 1^T	9	10		9	14	13	16	100	11	11
A. pomorum 3A^T									100	
Bacillus tusciae IFO 15312^T	7	9		6	5	7	5	4	7	100

Strain 1, *Alicyclobacillus* genomic species 1 DSM 11984; 2, *A. acidocaldarius* ATCC 27009^T; 3, *A. sendainensis* JNTAP-1^T; *4, A. acidoterrestris* ATCC 49025^T; 5, *A. acidiphilus* TA 67^T; 6, *A. hespericum* DSM 12489^T; 7, *A. cycloheptanicus* DSM 4006^T; 8, *A. herbarius* CP 1^T; 9, *A. pomorum* 3A^T; 10, *B. tusciae* IFO 15312^T.

3-3-6 Odor (guaiacol) production

Alicyclobacillus is not pathogenic bacteria [12]. However, some species produce an undesirable odor due to the production of guaiacol, causing deterioration of the quality of acidic beverages (especially those containing fruit juices). It has been known for some time that *A. acidoterrestris* produces guaiacol [37], but recently, *A. acidiphilus*, *A. herbarius* and *Alicyclo-*

bacillus sp. 2 (closely related to *A. hesperidum*: "*A. hesperidum* subsp. *aigle* " DSM 11985) were also reported as guaiacol producers (Table 3-7).

Table 3-6b. DNA base compositions and DNA-DNA hybridization values between isolates and type strains of related *Alicyclobacillus* species [89]

	Species	Strain	DNA G+C content (mol%)	1	2	3	4	5	6	7	8	9	10	11	12	13
	A. acidocaldarius	ATCC 27009^T	61.9	8	6	4	6	2	0	1	2	5	1	0	2	3
1	*A. saccharis*	RB718^T	56.6	100	12	6	11	5								
2	*A. hesperidum*	DSM 12489^T	56.2	10	100	7	8	5								
3	*A. fastidiosus*	S-TAB^T	53.9	6	4	100	13	4								
4	*A. acidoterrestris*	ATCC 49025^T	54.2	8	3	10	100	8	1	2	3	3	1	1	3	4
5	*A. acidiphilus*	TA67^T	57.5	6	2	5	10	100								
6	*A. kakegawaensis*	5-A83J^T	61.3						100	71	37	5	2			
7	*A. kakegawaensis*	5-A167N	61.7						74	100	37	8	2			
8	*A. shizuokaensis*	4-A336^T	60.5						34	30	100	7	2			
9	*A. herbarius*	CP1^T	59.6	4	2	4	6	1	6	5	5	100	3	6	4	4
10	*A. macrosporangidus*	5A-239-2O-A^T	62.5						2				100	5	4	3
11	*A. contaminans*	3-A191^T	60.6										2	100	75	6
12	*A. contaminans*	E-8	60.1										2	80	100	6
13	*A. pomorum*	3A^T	56.5										1	2	5	100
	A. cycloheptanicus	DSM 4006^T	60.6	4	1	3	8	2	0	2	2	2	2	3	3	3
	A. disulfidooxidans	DSM 12064^T	52.9						0	1	2	2	0	1	2	3

The strain *Alicyclobacillus* sp. 2 DSM 11985 is especially difficult to distinguish from *A. hesperidum*, due to a very close similarity even at the genomic level (Table 3-8). Concerning the relationship between *A. acidoterrestris* growth and the guaiacol production, it has been reported that production of guaiacol follows the growth curve, gradually accumulating and ultimately being degraded (Fig. 3-4) [11, 38-40]. Other species were not studied in detail. For detailed information about the mechanism of guaiacol biosynthesis, see section 3-6. Other off odor compounds are 2,6-dibromophenol and 2,6-dichlorophenol, these are also produced by *A. acidoterrestris* [41, 42]. There is a report that *A. cycloheptanicus* ATCC 49029 produces guaiacol in orange juice, but the criteria and taxonomic status of the strain have not yet been validated (this strain does not produce guaiacol from vanillic acid which is a precursor of guaiacol: our unpublished data).

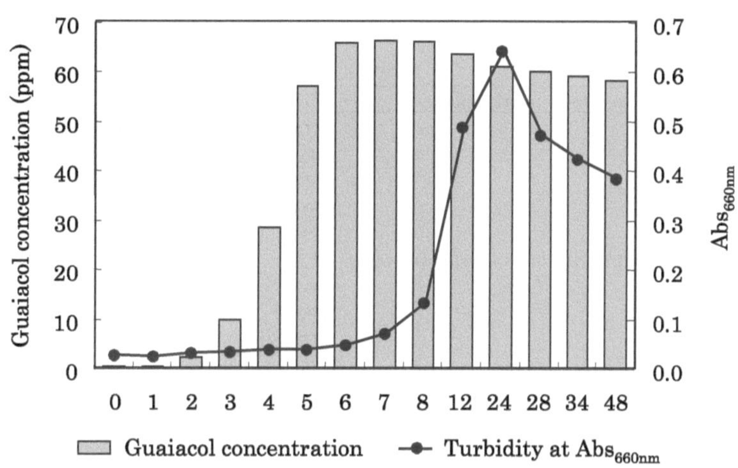

Fig. 3-4. Growth curve of *A. acidoterrestris* and guaiacol production in YSG-vanillic acid broth

A. acidoterrestris ATCC 49025^T spores suspension solution (5 ml, 2.0×10^7 CFU/ml) was inoculated into 195 ml of YSG liquid medium containing vanillic acid (final concentration: 100 ppm) and incubated at 45 °C for 48 h. Sampling was performed after appropriate intervals, and measured the guaiacol concentration by peroxidase method and the turbidity at Abs_{660nm}. As a reference, in the case of low spore concentration, the growth is slow and the production of guaiacol also delays (data not shown).

3-3-7 Oxygen requirement

It is well known that *Alicyclobacillus* is an obligate aerobic organism. Depending on culture conditions however, it has been reported that a very low level of oxygen, such as 0.1 % dissolved oxygen, can permit cell growth [43]. However, the growth ceases when oxygen is depleted, even if other nutrients are available. At this point the vegetative cells will sporulate.

Table 3-7. Guaiacol production ability of *Alicyclobacillus* strains

Species	Strain	Guaiacol (ppb)
Alicyclobacillus genomic species 1	DSM 11984	0
	DSM 11983	0
	DSM 453	0
	DSM 454	0
	DSM 12490	0
	KHA 31	0
	KHC 3	0
	MIH 2	0
	P2	0
	SO6	0
Alicyclobacillus genomic species 2	MIH 332	0
A. acidocaldarius subsp. *acidocaldarius*	ATCC 27009^T	0
	DSM 448	0
	DSM 449	0
	DSM 451	0
	DSM 452	0
	3B	0
	3W	0
	HP2	0
	MIH 321	0
	OJ5	0
	UZ 1	0
A. acidocaldarius subsp. *rittmannii*	DSM 11297^T	0
A. cycloheptanicus	DSM 4006^T	0
	DSM 4005	0
	DSM 4007	0
A. acidoterrestris	ATCC 49025^T	218
	DSM 2498	50
	DSM 3923	370
	DSM 3924	500
	B2065	40
	B-064	350
	OR3	120
A. acidiphilus	TA 67^T	480
A. hesperidum	DSM 12489^T	0
	DSM 12766	0
Alicyclobacillus sp. 1	isolate 3 (RB718)	0
Alicyclobacillus sp. 2	DSM 11985	146
	isolate 1	270
	isolate 2	295
	isolate 4	347
Alicyclobacillus sp. 3	8-E	0
A. herbarius	CP 1^T	490
A. pomorum	3A^T	0

Strains were inoculated into YSG liquid medium containing 1 ppm vanillin and cultured at each optimum temperatures for 72 hours. After cultivation, guaiacol concentration in the culture broth was analyzed by GC-MS. *Alicyclobacillus* sp. 4-8 do not produce guaiacol, but *Alicyclobacillus* sp. 9 produce guaiacol.

Table 3-8. Sequence differences of 16S rDNA and guaiacol production ability of *A. hesperidum* and the relatives

Species	Strain	Guaiacol production	Nucleotide position			
			98	131	148	149
A. hesperidum	DSM 12489^T	-	N	T	A	G
	DSM 12766	-	N	T	A	G
Alicyclobacillus sp. 1	isolate 3 (RB718)	-	A	C	G	G
Alicyclobacillus sp. 2	DSM 11985	+	A	C	A	A
	isolate 1	+	N	C	A	A
	isolate 2	+	N	C	A	A
	isolate 4	+	A	C	A	A

N: polymorphic site

Table 3-9. Effect of soluble solids on heat resistance of *Alicyclobacillus* WAC strain spores in concord grape juice [49)]

°Brix	Temp. (°C)	D-value (min)*	z-value
16	85	53.0	6.9 °C
	90	11.0	
	95	1.9	
30	85	76.0	6.6 °C
	90	18.0	
	95	2.3	
65	85	276.0	7.4 °C
	90	127.0	
	95	12.0	

*: Average of four trials.

3-3-8 Resistance to alcohol

Information about the alcohol (ethanol) tolerance of *Alicyclobacillus* is scarce. There is one report of a study that indicated growth was completely inhibited by the presence of 6 % alcohol (ethanol) [44)]. However, the spores were not affected by this alcohol concentration [45)].

3-3-9 Water activity

As with *Bacillus*, a high water activity (Aw>0.9) is needed for growth of *Alicyclobacillus*. In the case of fruit juices (depending on the type), it has been reported that a Brix close to 10 % allows cell growth, with an upper limit of 18-20 % Brix [19, 37, 45]. However, even in a Brix over 60 %, where there is no cell growth (for example concentrated fruit juice), the spores will survive. It has been reported that the heat resistance of *Alicyclobacillus* spores increases incrementally with increasing Brix concentration (Table 3-9) [46]. The relationship between *Alicyclobacillus* growth and juice type is discussed in more detail in Chapter 6.

Keiichi Goto

3-4 Heat resistance of *Alicyclobacillus* spores

As in the case of *Bacillus* and *Clostridium*, it is thought the temperature of the endospore formation, time (maturation), nutrients in the culture medium, and other factors affect the overall heat resistance of the spores of *Alicyclobacillus*. The reported D and z values differ considerably among researchers because the experimental conditions and protocols are not well established and the taxonomy of this group is still unclear. Basically, the heat resistance of *Alicyclobacillus* spores is thought to be lower than *Bacillus stearothermophilus*, *B. subtilis*, *Clostridium botulinum* and *C. tetani*. However, with the current pasteurization methods used for acidic beverages, there is the possibility spores will survive in the product and may cause subsequent spoilage. The heat resistance of the *Alicyclobacillus* spores is discussed below.

3-4-1 Heat resistance of *A. acidoterrestris* in commercial beverage products

In devising a strategy to prevent spoilage by *A. acidoterrestris*, it is important to investigate the heat resistance of *A. acidoterrestris* in commercially available beverages, and determine if destruction of the spores is possible under the conditions currently in use.

The heat resistance of *A. acidoterrestris* AB-1 strain, isolated from a spoiled beverage, was determined to be $D_{89°C}$=10.9-13.7 min and $D_{95°C}$=2.1-3.2 min (Table 3-10); the D value being the time required to reduce the microbial population to 1/10 of the initial number, under the testing temperature. This indicated that under the current thermal processing conditions (90-95°C, 15-20 sec) sterilization of the beverage is not sufficient [47].

Table 3-10. Heat resistance of *A. acidoterrestris* AB-1 strain spores in commercial beverages products [47]

Beverages	pH	°Brix	Temp.(°C)	D-value (min)*
Orange	3.8	11.8	89	13.7 ± 0.96
			95	3.2 ± 0.82
Fruit-blend	3.7	12.0	89	10.9 ± 0.97
			95	2.1 ± 0.20

*: Average of 5 samples±standard deviation.

3-4-2 Heat resistance of the *Alicyclobacillus* spores isolated from concentrated orange juice

The heat resistance of spores of *Alicyclobacillus* spp. isolated from concentrated orange juice (Strains No. 46, 70 and 145 strains) and tested in single strength orange juice were, respectively, $D_{85°C}$=60.8-94.5 min, $D_{90°C}$=10-20.6 min and $D_{95°C}$=2.5-8.7 min (Table 3-11) [48]. The z value is the temperature required to reduce the D value to 1/10 or increase 10 times. In the case of the strains shown in the Table 3-11, the range for the z value was 7.2-11.3°C. Therefore, 90-95°C, for 15-20 sec, the conditions currently being used, are not sufficient for sterilization of orange juice.

Table 3-11. Heat resistance of *Alicyclobacillus* spores isolated from concentrated orange juice [48]

Strain	D-value (min)			z-value
	85 °C	90 °C	95 °C	(°C)
No.46*	60.8	10.0	2.5	7.2
No.70*	67.3	15.6	8.7	11.3
No.145*	94.5	20.6	3.8	7.2
DSM2498	50.0	16.9	2.7	7.9

*:*Alicyclobacillus* sp.

3-4-3 Effect of soluble solids and pH on heat resistance

It is expected that the higher the soluble solids content (Brix), the greater the heat resistance will be, and the extent of this effect is discussed below. The effect of pH on heat resistance together with the content of soluble solids is also summarized there. The effect of pH on the heat resistance of *A. acidoterrestris* was compared under two different conditions: in an experimental model design (buffer) and in an actual fruit juices.

(1) Effect of soluble solids on heat resistance

When the heat resistance of spores of the *Alicyclobacillus* WAC strain, isolated from sterilized apple juice, was tested in Concord Grape Juice, it was found that the higher the soluble solids content, the greater the D values and the greater the heat resistance (Table 3-12) [49]. This indicated the destruction of spores present in concentrated juice is more difficult than in single strength juice.

Table 3-12. Effect of soluble solids on heat resistance of *Alicyclobacillus* WAC strain spores in concord grape juice [49]

°Brix	Temp. (°C)	D-value (min)*	z-value
	85	53.0	
16	90	11.0	6.9 °C
	95	1.9	
	85	76.0	
30	90	18.0	6.6 °C
	95	2.3	
	85	276.0	
65	90	127.0	7.4 °C
	95	12.0	

*:Average of four trials.

(2) Effect of soluble solids content and pH on heat resistance

Table 3-13 shows the effect of soluble solids content and pH on the heat resistance of the spores of *A. acidoterrestris* NCIMB 13137^T. Heat resistance increases as the soluble solids content increases and the D value increases at higher pH values [50].

Table 3-13. Effect of soluble solids and pH on heat resistance of *A. acidoterrestris* spores measured in malt extract broth [50)]

pH	SS* (°Brix or % by weight of sucrose)	Temp. (°C)	D±SE** (min)
2.5	5	85	35.5 ± 2.3
		97	0.771 ± 0.037
	60	85	60.3 ± 2.4
		97	2.15 ± 0.14
6	5	85	52.6 ± 1.6
		97	0.498 ± 0.045
	60	85	94.9 ± 6.7
		97	4.35 ± 0.27

*:Average of four trials.

(3) Effect of pH on heat resistance

(3)-1 Effect of pH on heat resistance in buffer solution

Spores of *A. acidoterrestris* AB-1 strain, isolated from a spoiled acidic beverage, were tested in citrate buffers (pH 2.5, 4.0, 6.0 and 6.9), and $D_{95°C}$ values of 0.9, 1.0, 0.9 and 0.9 min were obtained respectively, indicating there were no appreciable differences at different pH values [47)]. The AB-1 strain showed no significant differences in heat resistance at temperatures of 88-95°C in McIlvine buffer (citrate-phosphate buffer; pH 3.0-8.0) (Table 3-14), indicating the heat resistance of this spore is not pH-dependent [51)]. The effect of pH on the profiles of spore heat resistance of three strains of *A. acidoterrestris* (VF, WAC and IP strains) isolated from apple and other juices, showed that at 91°C the D value decreases as pH decreases, but this was not observed at 97°C. The resistance of spores of *A. acidoterrestris* VF strain at pH 3.1 in malic acid buffer, resulted in $D_{91°C}$ of 31.3 min, differing substantially from the result obtained at pH 3.7, which was $D_{91°C}$ of 54.3 min, with ($p \leq 0.05$). The $D_{97°C}$ at pH 3.1 was 7.9 min and at pH 3.7 was 8.8 min, not statistically different ($p > 0.05$, Table 3-15) [52)]. Therefore the results indicated that pH affect heat resistance of spore as same as temperature. Table 3-14 shows no apparent influence of pH on heat resistance; however in Table 3-15 some difference was detectable. One possible reason may be the differences between strains.

Table 3-14. Effect of pH on heat resistance of *A. acidoterrestris* AB-1 strain spores in McIlvine buffer [51)]

pH	Temperature			
	88 °C	90 °C	92 °C	95 °C
3.0	24.1 ± 1.63*[a]	14.8 ± 1.28[b]	6.2 ± 0.37[c, d]	2.7 ± 0.50[e]
4.0	25.9 ± 1.45[a]	16.1 ± 0.59[b]	6.1 ± 0.30[c, d]	2.8 ± 0.21[e]
5.0	29.1 ± 1.87[a]	16.6 ± 1.68[b]	7.1 ± 0.18[c]	2.7 ± 0.11[e]
6.0	25.9 ± 0.35[a]	16.8 ± 0.28[b]	6.8 ± 0.40[c, d]	2.3 ± 0.41[e]
7.0	24.7 ± 0.21[a]	15.7 ± 0.71[b]	6.7 ± 1.20[c, d]	2.2 ± 0.56[e]
8.0	25.7 ± 1.01[a]	16.1 ± 1.58[b]	5.7 ± 0.13[d]	2.3 ± 0.42[e]

*: Mean ± standard deviation.
a-e: Mean values with the same superscript are not significantly different at 5 % level.

Table 3-15. Effect of pH on heat resistance *A. acidoterrestris* VF, WAC and IP strain spores, in model systems containing 0.4 % malic acid [52)]

Strain	Temp.	pH	D-value (min)	z-value (°C)
Strain VF (spoiled apple juice)	91 °C	3.1	31.3	
		3.7	54.3	
	97 °C	3.1	7.9	10.0
		3.7	8.8	7.7
Strain WAC (nonspoiled apple-cranberry beverage)	91 °C	3.1	40.5	
		3.7	53.2	
	97 °C	3.1	8.0	8.5
		3.7	9.0	7.7
Strain IP (aseptically packged apple juice)	91 °C	3.1	20.3	
		3.7	32.6	
	97 °C	3.1	3.6	8.0
		3.7	3.8	6.5

Table 3-16. Effect of pH on heat resistance of *A. acidoterrestris* spores in grapefruit juice [53)]

Temp. (°C)	pH	D-value (min)*	z-value (°C)
80	3.0	31.85 ± 1.29	
	4.0	52.35 ± 1.94	
90	3.0	5.69 ± 0.21	
	4.0	9.44 ± 0.40	
95	3.0	1.49 ± 0.04	11.53
	4.0	1.73 ± 0.06	10.49

*: D-values shown are the average of two trials ± standard.

(3)-2 Effect of pH on heat resistance in fruit juice

The effect of pH on the heat resistance of *A. acidoterrestris* spores isolated from a spoiled drink containing fruit juice, was investigated using grape fruit juice as the heating medium. The pH affected heat resistance, with smaller D values observed at the lower pH (Table 3-16) [53].

3-4-4 Effect of the types of organic acid on heat resistance

The type of organic acid did not affect the heat resistance of the spores of three strains of *A. acidoterrestris* (VF, WAC and IP strains), isolated from fruit juices, such as apple, at either 91°C or 97°C (p>0.05, Table 3-17) [52].

Table 3-17. Effect of the types of organic acid on heat resistance of *A. acidoterrestris* VF strain spores [52]

Temp. (°C)	pH	Organic acid*	D-value (min)	z-value (°C)
91	3.1	malic acid	31.3	
		citric acid	46.1	
		tartaric acid	49.1	
	3.7	malic acid	54.3	
		citric acid	57.9	
		tartaric acid	69.5	
97	3.1	malic acid	7.9	10.0
		citric acid	8.2	8.5
		tartaric acid	8.4	7.8
	3.7	malic acid	8.8	7.7
		citric acid	10.8	8.2
		tartaric acid	10.0	7.1

*: malic; 0.40 %, citric; 0.58 %, tartaric: 0.45 % (range for commercial juices).

3-4-5 Effect of the types of fruit juice on heat resistance

Investigation of the heat resistance of *A. acidoterrestris* spores isolated from spoiled juice, found that orange juice, which has the highest pH, showed the highest D value at all temperatures evaluated (80-95°C). Grapefruit juice, which has the lowest pH, exhibited the lowest D value (Table 3-18) [53].

Table 3-18. Effect of the types of fruit juice on heat resistance of *A. acidoterrestris* spores [53]

Juice	pH	Temp. (°C)	D-value (min)*	z-value (°C)
		80	54.30 ± 0.42	
Orange	3.90	90	10.30 ± 0.30	12.9
		95	3.59 ± 0.04	
		80	41.15 ± 0.24	
Apple	3.51	90	7.38 ± 0.85	12.2
		95	2.30 ± 0.03	
		80	38.87 ± 0.20	
Grapefruit	3.42	90	5.95 ± 0.32	11.6
		95	1.85 ± 0.05	

*: the average of two trials ± standard deviation.

3-4-6 Effect of the heating medium on heat resistance

The heat resistance of *A. acidocaldarius* spores in citrate-phosphate buffer (pH 4.0 and pH 7.0), distilled water, and orange juice was determined. Results showed that at $D_{120°C}$=0.1 min and z value=7°C there were no significant differences between the tested heating media (p>0.05, Table 3-19) [54].

3-4-7 Effect of sporulating conditions on heat resistance

In order to prevent spoilage due to thermo-resistant microorganisms, it is necessary to correctly determine their thermo-tolerance. For this purpose, spores should be prepared to have high heat resistance as much as possible [55]. The genetically inherent resistance of the strain and the environmental conditions under which the spores were formed are thought to as well as affect their heat resistance. Other environmental factors include composition of the sporulation medium, incubation temperature at the sporulating phase [56], incubation time for spore maturation, metabolic products, and strain preservation conditions [57].

Table 3-19. Effect of the heating medium on heat resistance of *A. acidocaldarius* STCC 5137 strain spores [54)]

Heating medium	Temp. (°C)	D-value (min)*	z-value (°C)*
pH4 buffer**	110	2.6 ± 0.6	7.5 ± 2.4
	115	0.99 ± 0.43	
	120	0.11 ± 0.07	
	125	0.035 ± 0.002	
pH7 buffer**	110	2.6 ± 0.3	6.7 ± 1.0
	115	0.54 ± 0.07	
	120	0.097 ± 0.010	
	125	0.014 ± 0.002	
distilled water	110	3.7 ± 1.6	6.7 ± 0.3
	115	0.48 ± 0.17	
	120	0.11 ± 0.01	
	125	0.024 ± 0.007	
Orange juice	110	3.9 ± 0.1	6.8 ± 1.5
	115	0.61 ± 0.34	
	120	0.087 ± 0.009	
	125	0.027 ± 0.007	

*: Values are means ± 95% CL. **: McIlvine citrate-phosphate buffers.

(1) Effects of the sporulation medium composition on heat resistance

The addition of excessive amounts of manganese and calcium to the sporulating medium can increase the heat resistance [58)]. This information is summarized below:

(1)-1 Effects of $MnSO_4$ and citrate acid

The heat resistance in saline water of five different *A. acidoterrestris* strains isolated from spoiled acidic beverages, was the greatest with spores collected from cells grown in a medium containing 100 ppm $MnSO_4$ and 0.08 % citrate acid (Table 3-20). The observed resistance was $D_{95°C}$=1.3-2.2 min (z=5.9-6.9°C) for four of the five strains (TIF1701, TIF1702, TIF1703 and TIF1704 strains). For one strain (TIF1705 strain), the value was $D_{95°C}$=20.1 min (z=7.1°C) [55)].

(1)-2 Effect of divalent cations

Divalent cations (Ca^{2+}, Mg^{2+}, Ba^{2+}, Mn^{2+} and Sr^{2+}; 5 mmol/L each) have been reported to have no effect on the heat resistance of spores of *A. acidoterrestris* AB-1 strain, isolated from a spoiled beverage [59)].

The effect of Mn^{2+} in the culture medium on the heat resistance of spores of *A. acidoterrestris* was detected in (1)-1, however it was not detected in (1)-2. This fact may be due to the characteristics of the specific strains and the conditions of sporulation (medium, temperature and period of incubation time), etc.

Table 3-20. Effect of sporulation media on heat resistance *A. acidoterrestris* spores [55)]

Strain	Sporulation media (SMA)		D-value (min)		z-value
	Citric acid (%)	$MnSO_4$ (ppm)	90 °C	100 °C	(°C)
TIF 1701	0.04	0	6.1		
	0.04	100	9.2		
	0.08	0	3.2		
	0.08	100	8.8		6.0
TIF 1702	0.04	0	2.1		
	0.04	100	7.2		
	0.08	0	2.3		
	0.08	100	11.5		6.4
TIF 1703	0.04	0	3.5		
	0.04	100	10.5		
	0.08	0	4.2		
	0.08	100	11.8		5.9
TIF 1704	0.04	0	8.1		
	0.04	100	10.1		
	0.08	0	5.7		
	0.08	100	11.8		6.9
TIF 1705	0.04	0		-	
	0.04	100		2.1	
	0.08	0		1.6	
	0.08	100		3.5	7.1

(2) Effects of the temperature during spore formation on the heat resistance

Spores of *Bacillus* are known to become more resistant to heat treatment when the cells are exposed to higher temperatures during spore formation. In the same way, spores of *A. acidocaldarius* produced at 45°C had a resistance of $D_{110°C}$=0.48 min, while those produced at 65°C had a resistance of $D_{110°C}$=3.9 min, a value eight times higher [54)]. The same tendency was observed with *A. acidoterrestris*, where the heat resistance was lower in spores formed under lower temperatures [54)].

(3) Effects of the ion exchange treatment of the spore surface on heat resistance

Spores of *Bacillus subtilis* and *B. cereus* become less resistant to high temperatures when they are in H-form (acid treatment) as compared to those in N-form (no treatment). In some cases when H-form cells are changed to Ca-form (with an calcium ion treatment), we can observe an incremental change in heat resistance [59-62]. However, with *A. acidoterrestris* AB-1 strain isolated from a spoiled acidic beverage, there was no difference among the H-form, N-form and Ca-form spores. This would be due to strong bonding of Ca and Mn with the dipicolinic acid of the spores, which renders acid and calcium treatment ineffective for achieving ion exchange on the spore surface [59].

3-4-8 Compounds that decrease heat resistance

As discussed in 3-4-1, spores of *A. acidoterrestris* present in commercially available drinks cannot be eliminated using the current techniques of sterilization (90-95°C, 15-20 sec) [47], and so a way of reducing heat resistance of the spores is required. Nisin, lysozyme, and sucrose fatty acid esters (emulsifier) may act as agents for reducing the heat resistance of the spores. This is discussed below:

(1) Nisin

Nisin is a bacteriocin produced by *Lactococcus lactis* that has been used as a food preservative in 56 countries worldwide, with widespread use in Europe and America. It has not been approved for use in Japan, although this is currently under consideration. It is known that nisin has the ability to inhibit the growth of spores and vegetative cells of *A. acidoterrestris* [53, 63]. In apple juice, the heat resistance of *A. acidoterrestris* spores (a strain isolated from deteriorated fruit juice) was reduced by about 40 % with addition of nisin (50 IU/mL) (Table 3-21) [53].

(2) Lysozyme

Lysozyme has been shown to greatly reduce the heat resistance of *A. acidoterrestris* spores and this property has been confirmed in an application with acidic beverages (Table 3-22) [47]. Lysozyme is an enzyme responsible for the hydrolysis of β-1,4 bond between *N*-acetyl-glucosamine and *N*-acetyl-muramic acid of the peptidoglycans present in the bacterial cell wall. Due to the spore coat in the external layer of most of the spores, lysozyme cannot penetrate into the cortex consisting of petidoglycans, and is therefore not able to lyse the spore, but it can alter the permeability of the external layer and significantly reduce the heat resistance of the spore. The act of heating the enzyme can make it active against Gram-negative bacteria, which are not affected by lysozyme under normal conditions [64]. Heating may also change the stereostructure of the lysozyme molecule, making it permeable to the spore coat [47].

Table 3-21. Effect of nisin on heat resistance of *A. acidoterrestris* spores [53]

Nisin (IU/ml)	Temp. (°C)	D-value (min)*	z-value (°C)
	80	23.75 ± 0.12	
50	90	4.56 ± 0.09	13.8
	95	1.95 ± 0.02	
	80	41.15 ± 0.24	
0	90	7.38 ± 0.85	12.2
	95	2.30 ± 0.03	

*: Apple juice (pH 3.51), the average of two trials ± standard.

Table 3-22. Effect of Lysozyme on heat resistance of *A. acidoterrestris* AB-1 strain spores [47]

Heating medium	Lysozyme (%)	$D_{89°C}$ (min)
	0.000	10.8
	0.001	7.5
20mM Citric acid buffer (pH 4.0)	0.005	4.8
	0.010	4.1
	0.020	2.4
Commercial orange juice	0.000	13.7
	0.005	7.1

(3) Emulsifier

Sucrose fatty acid esters are emulsifiers that prevent germination of heat resistant spores and also exhibit effective antimicrobial activity against *Alicyclobacillus*. Sugar fatty acid esters composed of sucrose palmitate (monoester P: msP) added at 50 ppm has the ability to decrease the heat resistance (value of $D_{90°C}$) of *A. acidoterrestris* spores (Table 3-23) [65].

Table 3-23. Effect of sucrose palmitate (msP) on heat resistance of *A. acidoterrestris* DSM 3923 strain spores [65)]

Heating medium	msP (ppm)	$D_{90°C}$ (min)
Citric acid buffer	50	8.1
(pH 5.0)	0	9.7

3-4-9 Effect of aging on heat resistance

Spores of *A. acidoterrestris* NCIMB 13137^T, kept in frozen storage, showed an increase in z value as the spores aged (z value 7.8°C→29°C; Table 3-24) [66)]. Therefore, prior to juice sterilization, it is necessary to define the conditions of sterilization according to the intended period of frozen storage.

Table 3-24. Effect of aging on heat resistance of *A. acidoterrestris* NCIMB 13137 strain spores [66)]

Frozen storage (months)	$D_{95°C}$ (min)*	z-value (°C)
0	5.29 ± 0.96	7.8 ± 2.6
4	5.99 ± 0.63	22 ± 5
8	3.82 ± 0.48	29 ± 10

Takashi Tanaka

3-5 Toxicity and pathogenic characteristics of *Alicyclobacillus* and its metabolic products

Very little information is available on toxicity and the pathogenicity of *Alicyclobacillus*. In investigations carried out by Walls and Chuyate [67)], *A. acidoterrestris* and *A. acidocaldarius* cultured cells were injected, directly into the mouse stomach and in another experiment 1-3 mL of spoiled apple juice with turbidity and tainted odor (*A. acidoterrestris* 5×10^6 cell/mL) were fed by direct injection into the mouths of Guinea Pigs. No signs of pathogenicity were observed in either of these studies [67)]. To date, there have been no reports of any cases of illness attributed to *Alicyclobacillus*.

It is known that *Alicyclobacillus* produces compounds that are responsible for the off odors, and the toxicological data of these compounds is listed in Table 3-25 [68-70)]. As described in section 3-6-1, the guaiacol detected in juices with bad odor varies from approximately 10-200 ppb, and the 2,6-dibromophenol and 2,6-dichlorophenol were detected at levels of trace to 10 ng/L [71]. Many differences exist between humans and animals in physical constitution, assimilation by respiratory and digestive tracts, biochemical reactions, and other factors, therefore data obtained from animal experiments should be regarded only as an indication of what might occur in human subjects. However when we consider the data with respect to the general intake of soft drinks, it is reasoned that even if beverages contaminated by *Alicyclobacillus* are consumed, there is little risk of health problems since the consumption of odor related compounds is very low.

Table 3-25. Acute toxicity data of chemical substances related to off-flavor produced by *Alicyclobacillus* species [68-70]

Chemicals	Animal species	Route of exposure	Type of test	Dose
Guaiacol	rat	subcutaneous	LD_{Lo}	900 mg/kg
	mouse	intravenous	LD_{50}	170 mg/kg
	mouse	inhalation	LC_{50}	7570 mg/m^3
	rabbit	subcutaneous	LD_{Lo}	1250 mg/kg
	rabbit	skin	LD_{50}	4600 mg/kg
	guinea pig	subcutaneous	LD_{Lo}	900 mg/kg
	cat	oral	LD_{Lo}	1500 mg/kg
	rat	oral	LD_{50}	725 mg/kg
2,6-Dibromophenol	rat	oral	LD_{50}	≧2000 mg/kg
2,6-Dichlorophenol	rat	intraperitoneal	LD_{50}	390 mg/kg
	mouse	oral	LD_{50}	2120 mg/kg
2,4-Dichlorophenol	rat	subcutaneous	LD_{50}	1730 mg/kg
	rat	intraperitoneal	LD_{50}	430 mg/kg
	mouse	oral	LD_{50}	1276 mg/kg
	mouse	intraperitoneal	LD_{50}	153 mg/kg
2,4,6-Trichlorophenol	rat	intraperitoneal	LD_{50}	276 mg/kg
	rat	oral	LD_{50}	820 mg/kg
2,4-Dibromophenol	mouse	oral	LD_{50}	282 mg/kg
	rat	oral	LD_{50}	50 mg/kg

LDLo: Lowest Lethal Dose, The lowest dose of a substance introduced by any route, other than inhalation, reported to have caused death in humans or animals. LD50: Lethal Dose 50, a dose which is lethal to 50 % of the animals tested. LC50: Lethal Concentration 50, a concentration which is lethal to 50 % of the animals tested.

Rie Yamamoto, Teiichi Suzuki

3-6 Compounds responsible for off odor and their mechanisms of production

As previously discussed, in fruit juice contaminated with *Alicyclobacillus* there are occasional occurrences of distinctive "medicinal" or "smoky" odors which indicate a clear correlation between the presence of these bacteria and spoilage. The types of compounds that cause these odors and their respective mechanisms of production are discussed below.

3-6-1 Types and characteristics of off odor compounds

A. acidoterrestris is the main cause of off odors in fruit juices and is regarded as a serious cause of spoilage, for the fruit juice industry. The chemical compounds determined to be the cause of spoilage incidents and odor formation are guaiacol, 2,6-dibromophenol and 2,6-dichlorophenol [72-79]. Fig. 3-5 shows the structure of these compounds.

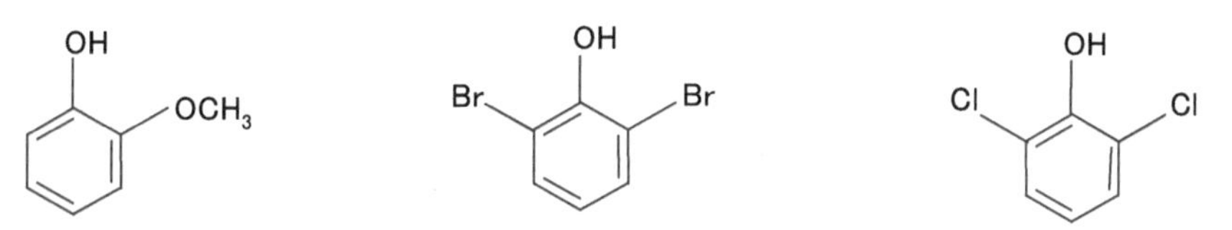

Guaiacol 2,6-Dibromophenol 2,6-Dichlorophenol

Fig. 3-5. Structures of odor compounds

Among these compounds, guaiacol is the primary and most significant concern. Guaiacol has an acidic and characteristic "medicinal" smell. It is possible to detect this odor at very low concentrations because the threshold of sensory detection is extremely low, 1 ppb in water and 10 ppb in orange juice. The amount of guaiacol detected in spoiled juice, varies from 10 to approximately 200 ppb, depending upon differences between strains and growth conditions. This is much higher than the sensory thresholds. When *A. acidoterrestris* is inoculated in orange juice and stored at 25°C, guaiacol is reported to be detected within a few days [80]. Based on the experience of this author, there are cases where growth could be observed within 1-2 days. Results obtained from an experiment carried out with *A. acidoterrestris*, indicated the generation of guaiacol was higher in orange than apple juice, and 46°C was more favorable than 37°C for guaiacol production [75]. The species that are known to produce guaiacol are: *A.*

Table 3-26. Growth of *Alicyclobacillus* bacteria and generation of guaiacol in acidic beverages [72]

	A. acidocaldarius ATCC 27009^T	"*A. mali*" TA 3	*A. acidoterrestris* DSM 3923	*A. acidiphilus* TA 67^T	*A. cycloheptanicus* DSM 4006^T	*A. herbarius* CP 1^T
Growth in acidic beverages	-	+	+	+	+	-
Generation of guaiacol from vanillin	-	-	+	+	-	+

acidoterrestris, *A. acidiphilus*, *A. herbarius* and *Alicyclobacillus* sp. 2 (closely related to *A. hesperidum* : "*A. hesperidum* subsp. *aigle* " DSM 11985). Among these species, *A. acidoterrestris* produces the highest amounts of guaiacol. There are few reports of spoilage caused by *A. acidiphilus* and *A. herbarius* [81]. The ability of the remaining species to produce guaiacol has not yet been observed. Table 3-26 summarizes the ability of several species of *Alicyclobacillus* to grow in acidic beverages, their growth profile, and guaiacol producing ability. As for the other species, *A. acidocaldarius*, "*A. mali* ", and *A. cycloheptanicus*, there are strains which can grow under experimental conditions, however the production of guaiacol has not been detected in any of them. On the other hand there are species such as *A. herbarius* which has the ability to produce guaiacol, yet growth has not been observed in beverages [72].

The growth phase in which guaiacol is produced is not always the same, and is affected by both the strain and temperature conditions. The variation in viable cell number of *A. acidoterrestris* and guaiacol production in orange juice is shown in Fig. 3-6. In this experiment, guaiacol was not produced during the active growth phase at temperatures of 35 or 55°C. However, guaiacol production was initiated when the cell number reached a steady state, or when the cell number started to decline [81].

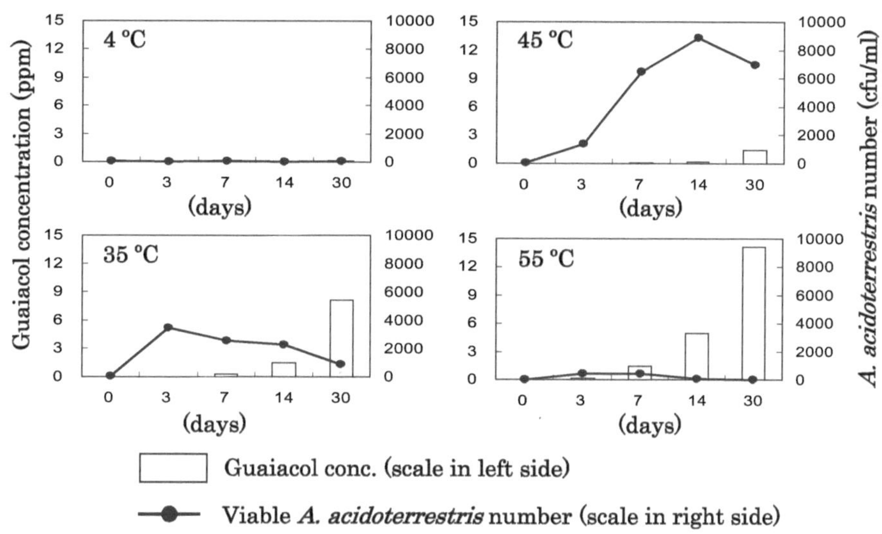

Fig. 3-6. Growth curve of *A. acidoterrestris* and guaiacol production in 100 % orange juice (Paper package) [81)]

2,6-Dibromphenol and 2,6-dichlorophenol are also, like guaiacol, undesirable odor compounds with a "medicinal" smell. The sensory detection thresholds of both compounds are 0.5 ng/L and 6.2 ng/L, respectively. The values for the bromophenol group are listed in Table 3-27 [71)]. The amount detected in fruit juices with off odor varied up to several tens ng/L, which is 1000 times lower than observed for guaiacol, but still at a concentration 10 times higher than the sensory limit. An example, of the detected values is shown in Table 3-28. In all cases, 2,6-dibromophenol and 2,6-dichlorophenol were detected over the limit of sensory detection [75)]. It has been reported that 2,6-dibromophenol was produced within 24 hours when a fruit juice was inoculated with *A. acidoterrestris* and stored in the presence of oxygen at 45°C [71)]. In addition to the two mentioned halophenols, 2,4-dibromophenol and 2,4-dichlorophenol and 2,4,6-trichlorophenol were also detected in odor tainted fruit juices, but their concentrations were below the sensory detection limit and were not directly responsible for the odor problem of the juice. While *A. acidoterrestis* is known to produce these odor-forming compounds, it has not been confirmed whether or not this is true in other species of *Alicyclobacillus*.

Table 3-27. Flavor threshold concentrations of bromophenols water and prawn meat [77)]

Compound	Water		Prawn meat	
	Threshold (μg/l)	Flavour description	Threshold (μg/kg)	Flavour description
2-Bromophenol	3×10^{-2}	Phenolic/iodine	2	Phenolic
4-Bromophenol	23	Phenolic	nd	nd
2,4-Dibromophenol	4	Phenolic	nd	nd
2,6-Dibromophenol	5×10^{-4}	Iodoform	6×10^{-2}	Phenolic
2,4,6-Tribromophenol	6×10^{-1}	Iodoform	nd	nd

nd: Not determined.

Table 3-28. Flavour threshold concentrations of bromophenols water and prawn meat [77)]

Sample	2,6-dibromophenol ng/L juice[a]	2,6-dichlorophenol ng/L juice[b]
Spoiled juice drink[c]	2	16
Inoculated juice drink[c], incubated for 23 days at 48 °C	14	40
Aust isolate 1 inoculated in apple juice incubated for 28 days at 46 °C	20	20
Aust isolate 1 inoculated in apple juice incubated for 28 days at 46 °C	20	20
German type strain inoculated in apple juice incubated for 28 days at 46 °C	10	20
Uninoculated juice, incubated and unincubated	ND[d]	ND[d]

a: Taste threshold 0.5 ng; 2,6-dibromophenol/L juice.
b: Taste threshold 30 ng; 2,6-dichlorophenol/L juice.
c: Pineapple, apple, passionfruit blend, 35 % juice.
d: Not detected. Limit of detection was around 1 ng halophenol/L juice.

3-6-2 Production mechanism of the odor compounds

Guaiacol has been investigated in detail regarding its production mechanism as an odor forming compound [72, 82, 83)]. As shown by the metabolic pathway in Fig. 3-7, guaiacol is produced by an oxidative reaction or by metabolism of ferulic acid, vanillin or vanillic acid. The conversion of ferulic acid to vanillin and vanillin to vanillic acid is mainly an oxidative process, although there are some indications that an enzymatic process is involved. Fig. 3-8 shows the oxidative pathway of vanillin from ferulic acid [83]. It could be surmised that the biosynthethic pathway from ferulic acid to vanillin has 4-vinyl-guaiacol as an intermediary compound, but there are some reports that there is no participation of 4-vinylguaiacol in the pathway [83]. Recently novel production mechanisms of guaiacol have been reported. It is proposed that in the formation of guaiacol from vanillin in orange juice, there is no contribution of the oxidative process and that the enzymatic process of *A. acidoterrestris* is the only dominant pathway [85]. Vanillin and vanillic acid are produced by many kinds of microorganisms such as bacteria, fungi and yeasts, however the production of guaiacol is observed with a very few microorganisms, mainly those from *Bacillus* and some Actinomycetes [82]. Within *Bacillus,* the bacteria known to produce guaiacol are: *Bacillus subtilis*, *B. megaterium*, *B. polymyxa* (formally *Paenibacillus polymyxa*) and *B. licheniformis*. Attention should be paid to the fact that once the conditions required by *Alicyclobacillus* to produce guaiacol are satisfied, other microorganisms may also cause spoilage [84].

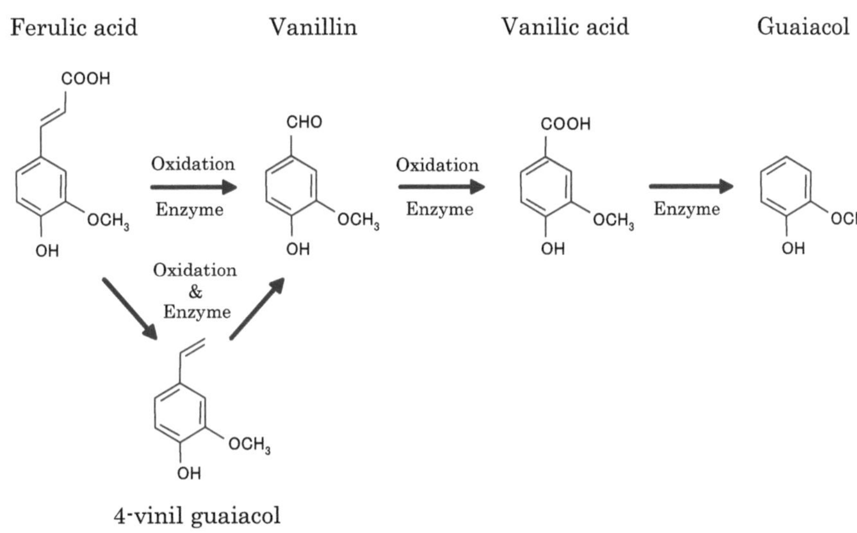

Fig. 3-7. Predictive pathway for the formation of guaiacol [73]

On the other hand, the mechanism of halophenol synthesis is not yet fully understood, although it is clear the key compounds in the biosynthetic pathway are the phenolics, hydrogen peroxide, halide ion and haloperoxidase. It has been suggested that, once *Alicyclobacillus* species possess an enzymatic system to promote halogenation of compounds, it becomes possible to convert trace amounts of phenolic compounds, hydrogen peroxide and halide ions present in fruit juices into halophenols [71].

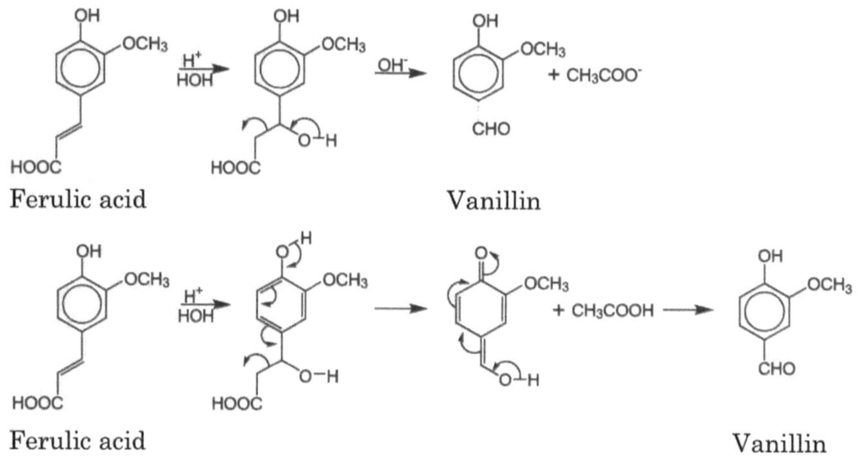

Fig. 3-8. Proposed mechanisms for the formation of vanillin from FA in MOJ solutions [83)]

Recently, in Japan, it was shown that, *in vitro* and in orange juice, ferulic acid and 4-vinylguaiacol are not the precursors of guaiacol, vanillic acid and vanillin, and also vanillic acid is converted from vanillin only by a bacterial enzyme which is produced by *A. acidoterrestris* [88)].

Hajime Tokuda

References

1. Uchino F, Doi S (1967) Acido-thermophilic bacteria from thermal waters. Agric Biol Chem 31:817-822.
2. Wisotzkey JD, Jurtshuk JRP, Fox GE, Deinhard G, Poralla K. (1992) Comparative sequence analyses on the 16S rRNA (rDNA) of *Bacillus acidocaldarius*, *Bacillus acidoterrestris*, and *Bacillus cycloheptanicus* and proposal for creation of a new genus, *Alicyclobacillus* gen. nov. Int J Syst Bacteriol 42:263-269.
3. Darland G, Brock TD (1971) *Bacillus acidocaldarius* sp. nov., an acidophilic thermophilic spore-forming bacterium. J Gen Microbiol 67:9-15.
4. DeRosa M, Gambacorta A, Minale L, Bu'Lock JD (1971) Cyclohexane fatty acids from a thermophilic bacterium. Chem Commun 21:1334a.
5. Deinhard G, Blanz P, Poralla K, Altan E (1987) *Bacillus acidoterrestris* sp. nov., a new themotolerant acidophile isolated from different soils. System Appl Microbiol 10:47-53.

6. Deinhard G, Saar J, Krischke W, Poralla K (1987) *Bacillus cycloheptanicus* sp. nov., a new thermoacidophile containing ω-cycloheptane fatty acids. System Appl Microbiol 10:68-73.
7. Wisse CA, Parish M (1998) Isolation and enumeration of sporeforming, thermo-acidophilic, rod-shaped bacteria from citrus processing environments. Dairy Food Environ Sanitation 18:504-509.
8. Eiroa MNU, Junqueira VCA, Schmidt F (1999) *Alicyclobacillus* in orange juice: occurrence and heat resistance of spores. J Food Protection 62:883-886.
9. Komitopoulou E, Boziaris IS, Davies EA, Delves-Broughton J, Adams MR (1999) *Alicyclobacillus acidoterrestris* in fruit juices and its control by nisin. Int J Food Sci Tech 34:81-85.
10. Duong HA, Jensen N (2000) Spoilage of iced tea by *Alicyclobacillus*. Food Australia 52:292.
11. Jensen N (2000) *Alicyclobacillus* in Australia. Food Australia 52:282-285.
12. Walls I, Chuyate R (2000) Spoilage of fruit juices by *Alicyclobacillus acidoterrestris*. Food Australia 52:286-288.
13. Walls I, Chuyate R (2000) Isolation of *Alicyclobacillus acidoterrestris* from fruit juices. J AOAC Int 83:1115-1120.
14. Silva FVM, Gibbs P (2001) *Alicyclobacillus acidoterrestris* spores in fruit products and design of pasteurization process. Trends Food Sci Tech 12:68-74.
15. Goto K, Matsubara H, Mochida K, Matsumura T, Hara Y, Niwa M, Yamasato K (2002) *Alicyclobacillus herbarius* sp. nov., a novel bacterium containing ω-cycloheptane fatty acids, isolated from herbal tea. Int J Syst Evol Microbiol 52:109-113.
16. Matsubara H, Goto K, Matsumura T, Mochida K, Iwaki M, Niwa M, Yamasato K (2002) *Alicyclobacillus acidiphilus* sp. nov., a new thermo-acidophilic ω-alicyclic fatty acid-containing bacterium isolated from acidic beverages. Int J Syst Evol Microbiol 52:1681-1685.
17. Cerny G, Hennlich W, Poralla K (1984) Fruchtsaftverderb durch Bacillen: Isolierung und Charakterisierung des Verderbserregers. Z Lebens Unters Forsch 179:224-227 (in German).
18. Suzuki T (1989) Sansei inryou kandume no kousankin ni tsuite. Shokuhin to Youki 30:503-506 (in Japanese).
19. Sprittstoesser DF, Churey JJ, Lee CY (1994) Growth characteristics of aciduric sporeforming bacilli isolated from fruit juices. J Food Protection 57:1080-1083.
20. Yamazaki K, Tezuka H, Shinano H (1996) Isolation and identification of *Alicyclobacillus acidoterrestris* from acid beverages. Biosci Biotech Biochem 60:543-545.
21. Kusano K, Niwa M, Yamasato H (1997) Kaju yori bunnri sareta tainetsusei kousanseikin no bunrui. Nihon-Seiryoinryo-Kenkyukai, Dai-6-kai Kenkyu Hapyokai (in Japanese).
22. Nicolaus B, Improta R, Manca CM, Lama L, Esposito E, Gambacorta A (1998) Alicyclobacilli from an unexplored geothermal soil in Antarctica: Mount Rittmann. Polar Biol 19:133-141.

23. Albuquerque L, Rainey FA, Chung AP, Sunna A, Nobre MF, Grote R, Antranikian G, DeCosta MS (2000) *Alicyclobacillus hesperidum* sp. nov. and a related genomic species from solfataric soils of Sao Miguel in the Azores. Int J Syst Evol Microbiol 50:451-457.
24. Goto K, Tanimoto Y, Tamura T, Mochida K, Arai D, Asahara M, Suzuki M, Tanaka H, Inagaki K (2002) Identification of thermoacidophilic bacteria and a new *Alicyclobacillus* genomic species isolated from acidic environments in Japan. Extremophiles 6:333-340.
25. Tsuruoka N, Isono Y, Shida O, Hemmi H, Nakayama T, Nishino T (2003) *Alicyclobacillus sendaiensis* sp. nov., a novel acidophilic, slightly thermophilic species isolated from soil in Sendai, Japan. Int J Syst Environ Microbiol 53:1081-1084.
26. Goto K, Mochida K, Asahara M, Suzuki M, Kasai H, Yokota A (2003) *Alicyclobacillus pomorum* sp. nov., a novel thermo-acidophilic, endospore-forming bacterium that does not possess ω-alicyclic fatty acids, and emended description of the genus *Alicyclobacillus*. Int J Syst Evol Microbiol 53:1537-1544.
27. Ryu E (1940) A simple method of differentiation between Gram-positive and Gram-negative organisms without staining. Kitasato Arch Exp Med 17:58.
28. Golovacheva RS, Karavaiko GI (1979) *Sulfobacillus* - a new genus of spore-forming thermophilic bacteria. Microbiology (English translation of Mikrobiologiya) 47:658-665.
29. Norris PR, Clark DA, Owen JP, Waterhouse S (1996) Characteristics of *Sulfobacillus acidophilus* sp. nov. and other moderately thermophilic mineral-sulphide-oxidizing bacteria. Microbiology 142:775-783.
30. Dufresne S, Bousquet J, Boissinot M, Guay R (1996) *Sulfobacillus disulfidooxidans* sp. nov., a new acidophilic, disulfide-oxidizing, gram-positive, spore-forming bacterium. Int J Syst Bacteriol 46:1056-1064.
31. Suzuki K, Saito K, Kawaguchi A, Okuda S, Komagata K (1981) Occurrence of ω-cyclohexyl fatty acids in *Curtobacterium pusillum* strains. J Gen Appl Microbiol 27:261-266.
32. Kusano K, Yamada H, Niwa M, Yamasato K (1997) *Propionibacterium cyclohexanicum* sp. nov., a new acid-tolerant ω-cyclohexyl fatty acid-containing propionibacterium isolated from spoiled orange juice. Int J Syst Bacteriol 47:825-831.
33. Kannenberg E, Blume A, Poralla K (1984) Properties of ω-cyclohexane fatty acids in membranes. FEBS 172:331-334.
34. Krischke W, Poralla K (1990) Properties of *Bacillus acidocaldarius* mutants deficient in ω-cyclohexyl fatty acid biosynthesis. Arch Microbiol 153:463-469.
35. Goto K, Mochida K, Asahara M, Suzuki M, Yokota A (2002) Application of the hypervariable region of the 16S rDNA sequence as an index for the rapid identification of species in the genus *Alicyclobacillus*. J Gen Appl Microbiol 48:243-250.
36. Wayne LG, Brenner DJ, Colwell RR, 9 other authors (1987) International Committee on Systematic Bacteriology. Report of the ad hoc committee on

reconciliation of approaches to bacterial systematics. Int J Syst Bacteriol 37:463-464.

37. Walls I, Chuyate R (1998) *Alicyclobacillus* - Historical perspective and preliminary characterization study. Dairy Food Environ Sanitation 18:499-503.
38. Pettipher GL, Osmundson ME, Murphy JM (1997) Method for the detection and enumeration of *Alicyclobacillus acidoterrestris* and investigation of growth and production of taint in fruit juices and fruit juice-containing drinks. Lett Appl Microbiol 24:185-189.
39. Orr RV, Shewfelt RL, Huang CJ, Tefera S, Beuchat LR (2000) Detection of guaiacol produced by *Alicyclobacillus acidoterrestris* in apple juice by sensory and chromatographic analysis, and comparison with spore and vegetative cell population. J Food Protection 63:1517-1522.
40. Niwa M (2003) *A. acidoterrestris* Rapid Detection Kit. Fruit Processing 13:328-331.
41. Borlinghaus A, Engel R (1997) *Alicyclobacillus* incidence in commercial apple juice concentrate (AJC) supplies - Method development and validation. Fruit Processing 7:262-266.
42. Jensen N, Whitfield FB (2003) Role of *Alicyclobacillus acidoterrestris* in the development of a disinfectant taint in shelf-stable fruit juice. Lett Appl Microbiol, 36:9-14.
43. Cerny G, Duong HA, Hennlich W, Miller S (2000) *Alicyclobacillus acidoterrestris*: influence of oxygen content on growth in fruit juices. Food Australia 52:289-291.
44. Splittstoesser DF, Churey JJ (1996) Unique spoilage organisms of musts and wines, p 36-41. In T. Toland and K C Fugelsang (ed), Wine spoilage microbiology Conference. California State University, Fresno.
45. Yamashita M (1996) Arukorurui no biseibutsu ni taisuru sayou. J Antibact Antifung Agents 24:25-49 (in Japanese).
46. Splittstoesser D F, Lee CY, Churey JJ (1998) Control of *Alicyclobacillus* in the juice industry. Dairy Food Environ Sanitation 18:585-587.
47. Yamazaki K, Isoda C, Tezuka H, Kawai H, Shinano H (1997) Sansei inryou henpai kin *Alicyclobacillus acidoterrestris* no tainetsusei to sono seigyo. Nihon Syokuhin Kagaku Kougakukaishi 44:905-911 (in Japanese).
48. Eiroa MNU, Junqueira VCA, Schmidt FL (1999) *Alicyclobacillus* in orange juice: Occurrence and heat resistance of spores. J Food Protection 62:883-886.
49. Splittstoesser DF, Lee CY, Churey JJ (1998) Control of *Alicyclobacillus* in the Juice Industry. Dairy Food Environ Sanitation 18:585-587.
50. Silva FM, Gibbs P, Vieira MC, Silva CLM (1999) Thermal inactivation of *Alicyclobacillus acidoterrestris* spores under different temperature, soluble solids and pH conditions for the design of fruit processes. Int J Food Microbiol 51:95-103.
51. Murakami M, Tedzuka H, Yamazaki K (1998) Thermal resistance of *Alicyclobacillus acidoterrestris* spores in different buffers and pH. Food Microbiol 15:577-582.

52. Pontius AJ, Rushing JE, Foegeding PM (1998) Heat resistance of *Alicyclobacillus acidoterrestris* spores as affected by various pH values and organic acids. J Food Protection 61:41-46.
53. Komitopoulou E, Boziaris IS, Davies EA, Delves-Broughton J, Adams MR (1999) *Alicyclobacillus acidoterrestris* in fruit juices and its control by nisin. Int J Food Sci Tech 34:81-85.
54. Palop A, Alvarez I, Raso J, Condo'n S (2000) Heat resistance of *Alicyclobacillus acidocaldarius* in water, various buffers, and orange juice. J Food Protection 63:1377-1380.
55. Matsui T, Endo M, Nakao H, Yoshimoto S (1998) Kousansei kin *Alicyclobacillus acidoterrestris* no tainetsusei ni tsuite. Toyo Shokuhin Kogyo College Toyo-Shokuhin Institue Kenkyu Hokokusyo 22:123-127 (in Japanese).
56. Lechowich RYV, Ordal ZJ (1962) The influence of the sporulation temperature on the heat resistance and chemical composition of bacterial spores. Can J Microbiol 8:287-295.
57. Gorman SP, Scott EM, Hutchinson EP (1985) Thermal resistance variations due to post-harvest treatments in *Bacillus subtilis* spores. J Appl Bacteriol 59:555-560.
58. Amaha M, Ordal ZJ (1957) Effect of divalent cations in the sporulation medium on the thermal death rate of *Bacillus coagulans* var. *thermoacidurans*. J Bacteriol, 74:596-604.
59. Yamazaki K, Kawai Y, Inoue N, Shinano H (1997) Influence of sporulation medium and divalent ions on the heat resistance of *Alicyclobacillus acidoterrestris* spores. Lett Appl Microbiol, 25:153-156.
60. Raso J, Palop A, Bayarte M, Condo'n S, Sala FJ (1995) Influence of sporulation temperature on the heat resistance of a strain of *Bacillus licheniformis*. Food Microbiol 12:357-361.
61. Melly E, Genest PC, Gilmore ME, Little S, Popham DL, Driks A, Setlow P (2002) Analysis of the properties of spores of *Bacillus subtilis* prepared at different temperatures. J Appl Microbiol 92:1105-1115.
62. Kovacs-Proszt G, Farks J (1976) Acta Alimentaria 5:179-188.
63. Yamazaki K, Murakami M, Kawai Y, Inoue N, Matsuda T (2000) Use of nisin for inhibition of *Alicyclobacillus acidoterrestris* in acidic drinks. Food Microbiol, 17 315-320.
64. Azumaguchi S, Ibrahim HR, Juneja LR, Fujiki Y, Kim T, Yamamoto T (1996) Netsushori niyoru ranpaku lizochiumu no koukinsei no kaihen. Nihon Nougei Kagaku Kai Taikai 1996 nendo taikai kouen youshishu (in Japanese).
65. Inryou ni okeru shotou shibousan esuteru no kousansei gahoukin yokusei kouka. Ryoto® Mono Esuteru-P Shiryou. Mitsubishi Kagaku Fuzu Co., Ltd. (in Japanese).
66. Vieira MC, Teixeira AA, Silva FM, Gaspar N, Silva CLM (2002) *Alicyclobacillus acidoterrestris* spores as a target for Cupuacu nectar thermal processing: kinetic parameters and experimental methods. Int J Food Microbiol 77:71-81.

67. Walls I, Chuyate R (2000) Spoilage of fruit juices by *Alicyclobacillus acidoterrestris*. Food Australia 52:286-288.
68. Kagaku busshitsu anzen jyouhou teikyou shisutemu (KIS-NET), Kanagawaken Kankyou-Kagaku Senta, http://www.k-erc.pref.kanagawa.jp/kisnet/ (in Japanese).
69. Kagaku busshitsu sougou deta besu, Chikyukibo-Kagakubusshitsu Jyouhou Nettowaku. http://wwwdb.mhlw.go.jp/ginc/html/db1-j.html (in Japanese).
70. SIRI (Safety Information Resouces,Inc) MSDS Index. http://hazard.com/msds/index.php
71. Jansen N, Whitfield FB (2003) Role of *Alicyclobacillus acidoterrestris* in the development of a disinfectant taint in shelf-stable fruit juice. Lett Appl Microbiol 36:9-14.
72. Goto K (2000) Kouonsei kousansei gahou keisei saikin: *Alicyclobacillus* zoku saikin. J Antibact Antifung Agents 28:499-508 (in Japanese).
73. Goto K (2003) Tainetsusei kousansei kin (*Alicyclobacillus* zoku saikin) no kensa kanbetsuhou ni tsuite. Nihon Kaju-Kyokai (in Japanese).
74. Eguchi SY, Gilson P, Marcia M, Pinhatti EC, Azuma EH, Variane S (1999) Colecao de Culturas Tropical (CCT) Fundacao Andre Tosello and ABECitrus, Report of Research Project : Acidotermophilic sporeforming bactéria (ATSB) in orange juices. Detection methods, ecology and involvement in the deterioration of fruit juices. (in Japanese and Protégés).
75. Jensen N (2000) *Alicyclobacillus* in Australia. Foods Australia 52:282-285.
76. Jensen N (1999) *Alicyclobacillus* - a new challenge for the food industry. Foods Australia 51:33-36.
77. Whitfield FB, Last JH, Shaw KJ, Tindale CR (1988) 2,6-Dibromophenol: The cause of an iodoform-like off-flavour in some Australian crustacea. J Sci Food Agric 46:29-42.
78. Borlinghaus A, Engel R (1997) *Alicyclobacillus* incidence in commercial apple juice concentrate (AJC) suppliers - Method development and validation. Fruit Processing 7/97:262-266.
79. Yamazaki K, Teduka H, Shinano H (1996) Isolation and identification of *Alicyclobacillus acidoterrestris* from acidic beverages. Biosci Biotech Biochem 60:543-545.
80. Pettipher GL, Osmundson ME, Murphy JM (1997) Methods for the detection and enumeration of *Alicyclobacillus acidoterrestris* and investigation of growth and production of taint in fruit juice and juice-containing drinks. Lett Appl Microbiol 25:185-189.
81. Goto K (2003) Saishin tainetsusei kousansei kin no kiso chishiki to sono taisakuhou. Tainetsusei kousansei kin shinpojiumu ILSI JAPAN (in Japanese).
82. Rosazza JPN, Huang Z, Dostal L, Volm T, Rousseau B (1995) Review: Biocatalytic transformations of ferulic acid: an abundant aromatic natural product. J Industrial Microbiol 15:457-471.
83. Peleg H, Naim M, Zehavi J, Rouseff RL, Nagy S (1992) Pathways of 4-vinylguaiacol formation from ferulic acid in model solutions of orange juice. J Agric Food Chem 40:764-767.

84. Niwa M (2004) Guaiakoru kensyutu kitto (Peruokishi daze hou) niyoru sansei inryou yuugai kin no kanri. Japan Food Sci 43:23-28 (in Japanese).
85. Goto K (2005) Orenji jusu niokeru guaiakoru seisei ni itarumadeno jidousanka to biseibutukoso no kanyo. Heisei 17nendo Kaju gijutu kenkyu happyokai (in Japanese).
86. Karavaiko GI, Bogdanova TI, Tourova TP, Kondrat'eva TF, Tsaplina IA, Egorova MA, Krasil'nikova EN, Zakharchuk LM (2005) Reclassification of '*Sulfobacillus thermosulfidooxidans* subsp. *thermotolerans*' strain K1 as *Alicyclobacillus tolerans* sp. nov. and *Sulfobacillus disulfidooxidans* Dufresne et al. 1996 as *Alicyclobacillus disulfidooxidans* comb. nov., and emended description of the genus *Alicyclobacillus*. Int J Syst Evol Microbiol 55:941-947.
87. Simbahan J, Drijber R, Blum P (2004) *Alicyclobacillus vulcanalis* sp. nov., a thermophilic, acidophilic bacterium isolated from Coso Hot Spring, California, USA. Int J Syst Evol Microbiol 54:1703-1707.
88. Goto K, Fujita R, Asahara M, Oono A, Koda A, Suzuki M (2005) Orenji jusu ni okeru guaiakoru seisei ni itaru madeno jidousanka to biseibutu kouso no kanyo. Nihon Kaju-Kyokai Gijutu Kenkyu Kai Youshishu (in Japanese).
89. Goto K, Mochida K, Kato Y, Asahara M, Fujita R, An SY, Kasai K, Yokota A Proposal of *Alicyclobacillus contaminans* sp. nov., *Alicyclobacillus fastidiosus* sp. nov., *Alicyclobacillus kakegawaensis* sp. nov., *Alicyclobacillus macrosporangidus* sp. nov., *Alicyclobacillus sacchari* sp. nov., and *Alicyclobacillus shizuokaensis* sp. nov., novel moderately-thermophilic, acidophilic, endospore-forming bacteria. Int J Syst Evol Microbiol (in review).
90. Simbahan J, Drijber R, Blum P (2004) *Alicyclobacillus vulcanalis* sp. nov., a thermophilic, acidophilic bacterium isolated from Coso Hot Spring, California, USA. Int J Syst Evol Microbiol 54:1703-1707.
91. Karavaiko GI, Bogdanova TI, Tourova TP, Kondrat'eva TF, Tsaplina IA, Egorova MA, Krasil'nikova EN, Zakharchuk LM (2005) Reclassification of '*Sulfobacillus thermosulfidooxidans* subsp. *thermotolerans*' strain K1 as *Alicyclobacillus tolerans* sp. nov. and *Sulfobacillus disulfidooxidans* Dufresne et al. 1996 as *Alicyclobacillus disulfidooxidans* comb. nov., and emended description of the genus *Alicyclobacillus*. Int J Syst Evol Microbiol 55:941-947.
92. Goto K, Mochida K, Kato Y, Asahara M, Ozawa C, Kasai H, Yokota A (2006) Diversity of *Alicyclobacillus* isolated from fruit juices and their raw materials, and emended description of *Alicyclobacillus acidocaldarius*. Microbiol Cult Coll 22:1-14.

Chapter 4: Parameters for detection of *Alicyclobacillus* and test methods

4-1 Background

Since 1982 when incidents of deterioration due to *Alicyclobacillus* (later identified as *A. acidoterrestris*) occurred in Germany [1], more than 40 different test methods to detect *Alicyclobacillus* have been developed worldwide. This can partly be attributed to the thermophilic and acidophilic characteristics of *Alicyclobacillus* which make *in vitro* cultivation of the bacteria so difficult. Sensitivity and precision varies among the methods. Differences between the test results obtained in the country of production and those obtained after the importation of the product, has generated commercial conflicts. This situation is also influenced by differences among countries in perceptions regarding thermo-acidophilic bacteria and the commercial damage caused by them. In 2001, the IFU (International Federation of Fruit Juice Producers, Paris, France), began international cooperative efforts towards standardization of the methods for detecting thermo-acidophilic bacteria (*Alicyclobacillus*). These efforts resulted in publication of the "First Standard IFU-Method on the Detection *Alicyclobacillus* in Fruit Juices" in April 2003. Later, a revision of the IFU method, entitled "Method on the Detection of taint producing *Alicyclobacillus* in Fruit Juices, revision 1" was published in September 2004 (IFU Handbook Microbiological Methods, Method No.12).

In Japan on November 6, 2002, the Japan Fruit Juice Association (JFJA) organized a joint meeting with beverage manufacturing companies (ten in total) to discuss the development and adoption by the industry of a unified method for detection of "Thermo-acidophilic bacteria (*Alicyclobacillus*)". Participants discussed factors conducive to contamination of foodstuffs by *Alicyclobacillus*, and efforts to detect and control the bacteria They agreed to adopt a unified method developed by a Working Group of the JFJA, which started its activities in Spring, 2003. After many meetings and the collection of a large amount of experimental data, guidelines for investigating organisms belonging to the genus *Alicyclobacillus* were proposed. Each of the participating companies then independently validated the guidelines. The results showed that there was either no difference, or an improvement in the sensitivity of the suggested methodology compared with the one normally used by each company. After that, on February 28th 2003, the "Unified method for detection of thermo-acidophilic bacteria" (see 4-3) was promulgated [details are available in the Bulletin of the Juice Producers Association, March edition (Japanese version)]. This method was presented at the IFU - Microbiology Working Group (April 9th, 2003, Köln, Germany) with the aim of reaching an international alignment. Improvements to the approved methodology were added by this Working Group in the Spring of 2004, and suggestions for an even simpler and more accurate method resulted in a revision [see "4-3 Unified Method for Detection of Thermo-acidophilic bacteria (4)"]. This revised method was also presented at the IFU Symposium 2004 (Stuttgart, Germany), and the Working Group is presently engaged in disseminating information on this topic in Japan and other countries. The method was published in English on February 23th, 2003 (details are available in the Bulletin of the Juice Producers Association) [15)]. It is hoped that if the *Alicyclobacillus* detection methods are unified in Japan, that in time they will also be accepted and implemented internationally.

This book compiles the parameters which affect the detection of *Alicyclobacillus* (including some unpublished data), and introduces the "Unified method for detection of thermo-acidophilic bacteria" developed by the JFJA, as well as development work carried out by the IFU on the "Method on the Detection of taint producing *Alicyclobacillus* in Fruit Juices, revision 1". The present guidelines for *Aliyiclobacillus* detection are also discussed.

4-2 Parameters that affect the detection of *Alicyclobacillus*

In addition to the usual parameters of cultivation such as medium type, temperature, growth period, pH and heat shock, the type of fruit and the concentration of the juice play an essential role in *Alicyclobacillus* detection. Based on the publications available on this subject, each of these parameters will be reviewed and discussed further. The information was verified by the JFJA, the IFU and ABECitrus (Brazilian Association for Citrus Export).

4-2-1 Growth medium

At present, YSG and BAT media are the most frequently utilized growth media for the detection of *Alicyclobacillus*, followed by the Modified Plate Count Agar (=mPCA), OSA, PDA, K-Agar, MEA, BAM and Semi-synthetic medium. Less frequently utilized media include BCY, BSSA, AAM, TYG, ALI, DTA and Modified YPGA. The geographical distribution of the use of these media is as follows: YSG is used in Japan and Brazil, BAT in Europe, and mPCA, OSA, PDA and K-Agar in the United States and Australia. Comparative performance data for different media has appeared in a number of publications. However, there is a consensus that YSG and BAT media are more sensitive for the detection of *Alicyclobacillus* growth. The composition of YSG medium is very simple, however YSG medium does not have buffering capacity and therefore is not appropriate for long term preservation of *Alicyclobacillus* strains because acidification of the medium occurs. The following are examples of the comparative performance of the main media types used worldwide.

(1) Walls (now USDA Food Safety and Inspection Service) and Chuyate from ILSI Risk Science Institute tested the capacity of K-agar, Semi-synthetic medium and OSA to recover *A. acidoterrestris* spores using the Pour-Plate Technique [3, 4)]. From this study, K-agar showed the best recovery capacity in all types of juices tested.

(Addendum) K-agar is a medium developed by Kirin Beverage Co., Ltd. (Japan) being an abbreviation of Kirin-agar. It was observed that Tween-80 in K-agar inhibited growth of some strains of *A. acidoterrestris*. Kirin Beverage Co., Ltd. subsequently developed YSG medium with a focus on improving sensitivity to all *Alicyclobacillus* strains. Recent data indicate that K-agar is also suitable for selective detection of *A. acidoterrestris*.

(2) Fujita and Goto, members of the JFJA Working Group, evaluated the performance of BAT agar and YSG agar (unpublished data). Results of this work, which were included in the Working Group Technical Report of 2003, are summarized as follows. Using the "Surface Plating Technique" no difference in growth of vegetative cells of *Alicyclobacillus* spp. was observed between the media. Using the "Membrane Filter Technique" there was almost no difference when comparing media for the recovery of spores of *Alicyclobacillus* spp. using the "Membrane Filter Technique" (Table 4-1 and Fig 4-1). However, when the "Pour Plate Technique" was used for the recovery of spores of *A. acidoterrestris*, twice as many spores were recovered on YSG agar than on BAT agar (Table 4-2). A comparison

between Difco-base YSG agar (Bacto^{TM} yeast extract: Code No. 212750, soluble starch: Code No. 217820, Bacto^{TM} agar: Code No. 214010) and Merck-base YSG agar (yeast extract: Code No. 1.11926.1000, soluble starch: Code No. 1.01252.0100, agar: Code No. 1.11925.1000) showed no significant difference between these agars when vegetative cells were cultured using the "Surface Plating" or "Membrane Filter Technique". When growth was compared using the "Pour Plating Method", a slightly higher number of colonies grew on the Merck-base YSG agar (Table 4-3).

Table 4-1. Growth of *Alicyclobacillus* species on BAT agar and YSG agar

Species	Strain	BAT agar	YSG agar
Alicyclobacillus genomic species 1	DSM 6481	Good	Good
	DSM 11983	Good	Good
	SO6	Good	Good
	TA3	Good	Good
Alicyclobacillus genomic species 2	MIH332	Good	Good
A. acidocaldarius	ATCC 27009^{T}	Good	Good
	OJ5	Good	Good
	MIH321	Good	Good
	3W	Weak	Good
A. acidoterrestris	ATCC 49025^{T}	Good	Good
	DSM 2498	Good	Good
	DSM 3923	Good	Good
	DSM 3924	Good	Good
	OR3	Good	Good
	B2065	Good	Good
	B2066	Good	Good
	B2067	Good	Good
	RB1	Good	Good
	RB221	Good	Good
	RB253	Good	Good
	RB346	Good	Good
	RB359	Good	Good
	TAB-H1	Good	Good
A. acidiphilus	TA67^{T}	Good	Good
A. herbarius	CP1^{T}	Good	Weak
A. hesperidum	DSM 12489^{T}	Good	Good
A. cycloheptanicus	DSM 4006^{T}	Weak	Weak
A. pomorum	3A^{T}	Good	Good

Growth was observed after 48 h incubation. Strains of *A. acidiphilus*, *A. acidoterrestris*, *A. cycloheptanicus*, *A. hesperidum* and *A. pomorum* were cultured at 50 °C. Strains of *Alicyclobacillus* genomic species 1, *Alicyclobacillus* genomic species 2, *A. acidocaldarius* and *A. herbarius* were cultured at 60 °C.

Table 4-2. Detection of *A. acidoterrestris* spore by pour plating method

(CFU/plate)

Strain	BAT agar			YSG agar		
	plate 1	plate 2	average	plate 1	plate 2	average
ATCC 49025^T	15	16	16	25	34	30
B2065	12	12	12	52	28	40
B2066	20	28	24	69	53	61
B2067	39	40	40	47	51	49
RB1	20	20	20	31	36	34
RB221	8	10	9	64	48	56
RB253	13	14	14	47	55	51
RB346	26	30	28	62	70	66
RB359	9	20	15	34	41	38
TAB-H1	30	38	34	34	41	38

Fifty *A. acidoterrestris* spores were inoculated into 20 ml of BAT agar or YSG agar using the pour plating method. After incubation at 45 °C for 5 days, cfu were determined.

Table 4-3. Detection of the spores of eight *A. acidoterrestris* strains in Difco-based and Merck-based YSG agar

Strain	Difco	Merck
M-1	4.35×10^7	3.14×10^7
M-2	8.70×10^5	9.05×10^5
M-3	1.38×10^6	2.05×10^6
M-4	4.40×10^6	8.45×10^6
M-5	5.05×10^7	4.70×10^7
M-6	7.65×10^7	6.40×10^7
M-7	4.60×10^6	1.21×10^7
M-8	1.41×10^7	1.75×10^7

Strains of *A. acidoterrestris* were cultured at 45 °C for 7 days then spores were collected by centrifigation (3000 rpm × 10 min, twice). After heat shock treatment at 70 °C for 20 min, the spore suspension was inoculated into Difco-based and Merck-based YSG agar using the pour plating method. After incubation at 45 °C for 5 days, cfu were determined.

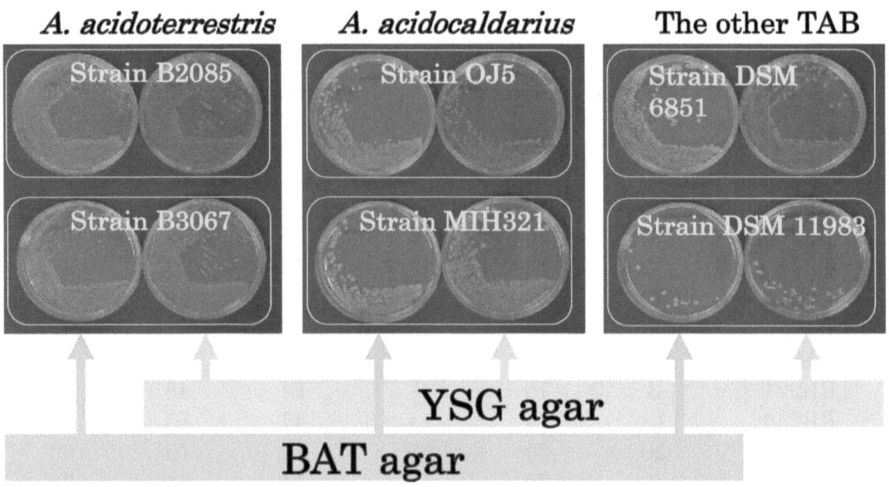

Fig. 4-1. Comparison of BAT agar and YSG agar for the growth of *A. acidoterrestris*

Each strain was spread on BAT agar and YSG agar and incubated at optimum temperatures for 24 hours.

(3) Members of the Technical Committee of ABECitrus have tested BAT agar and YSG agar with *Alicyclobacillus* isolates, using initial spore concentrations in orange juice of 10^1-10^4 cfu/mL [5)]. YSG agar was clearly more effective for spore recovery at the 10^1 cfu/mL level (Fig. 4-2), but at higher levels of inoculum there was no difference between the two media (Fig. 4-3). In growth promotion tests, YSG agar was more effective in the first 3 days of cultivation; however, for longer periods there was no difference between the media. In summary, YSG agar was the most effective in regards to sensitivity, but it may have limited species selectivity.

(Addendum 1) The selectivity of the YSG medium (both solid and liquid) is limited, because it also allows the growth of aerobic spores from acidophilic or acido-tolerant bacteria such as *B. subtilis* and/or *B. fumarioli.*

(Addendum 2) YSG and BAT media are commercially available from Merck under the codes: Code No. 1.07206.0500 (YSG agar), Code No. 1.07207.0500 (YSG broth) and Code No. 1.07904.0500 (BAT medium). BBL distributes YSG agar ready medium (Code No. 251841).

(Addendum 3) YSG medium is recommended by the JFJA, while K-agar in combination with BAT or YSG medium is recommended by IFU.

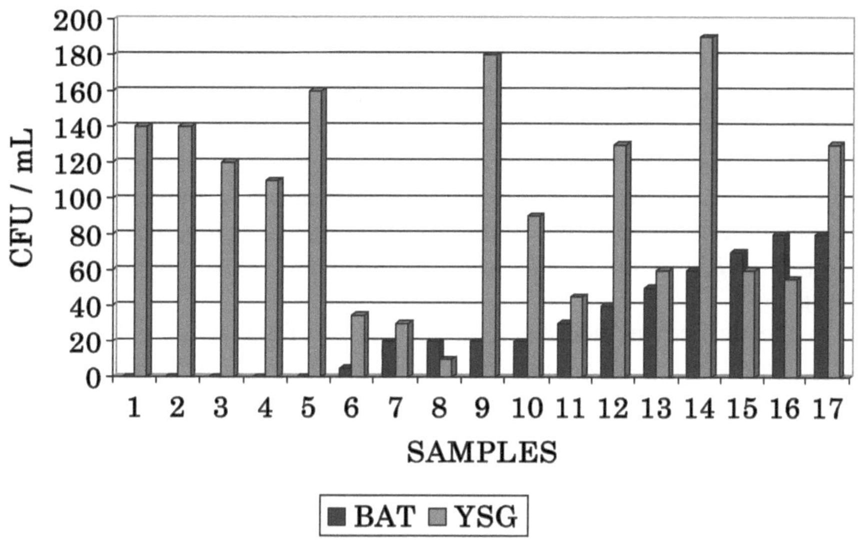

Fig. 4-2. Comparison of *A. acidoterrestis* spores recovery from BAT and YSG media (inoculation level: 10 CFU/ml)

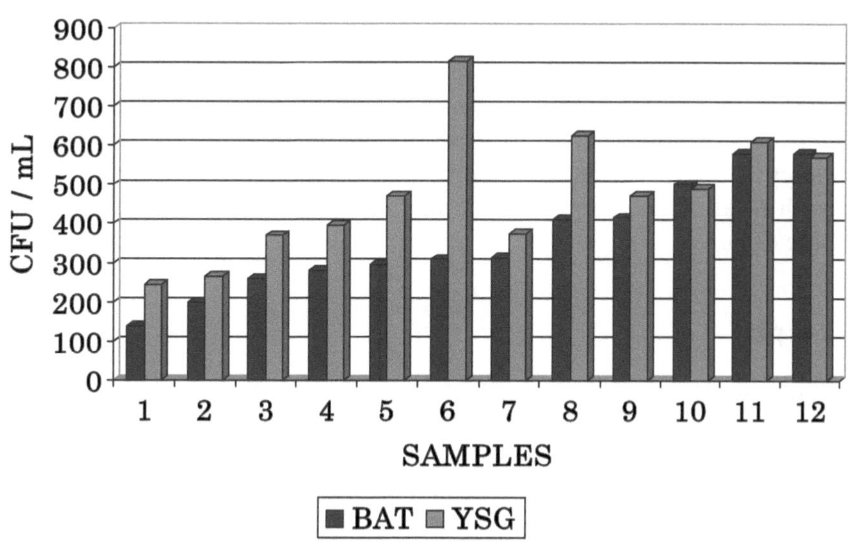

Fig. 4-3. Comparison of *A. acidoterrestis* spores recovery from BAT and YSG media (inoculation level: 100 CFU/ml)

4-2-2 Temperature for growth

The temperature range and optimum pH for the growth of *Alicyclobacillus*, grown in YSG broth are as shown in Table 4-4 [6, 7]. These values differ among strains (Table 4-5 and Fig. 4-4). The growth temperature used in the "Unified Method for Detection of Thermo-acidophilic bacteria" is 45°C, but this is the temperature that was determined to be best for detection of *A. acidoterrestris*, the representative guaiacol-producing species. A temperature of 50-55°C is suitable for detecting *A. acidocaldarius* and the *Alicyclobacillus* genomic species 1 (Mitsuda and co-workers, unpublished data).

In the unified methodology, growth is on YSG agar at 30°C. Under these conditions, *A. acidoterrestris*, *A. acidiphilus*, *A. cycloheptanicus* and *A. hesperidum* form colonies within 5 days of incubation whilst other species do not (JFJA Working Group, unpublished data). Incubation at 45°C allows the growth of all species.

4-2-3 Incubation period

Satisfactory growth is observed when a large amount of fresh vegetative cells (< 1000 cell/ml) are inoculated into YSG liquid medium, and incubated at the optimal growth temperature for 24 h. However, when the inoculum concentration is very low (> 1 cell/ml), growth will not necessarily be detected after 24 h. Therefore the "Unified method for detection of thermo-acidophilic bacteria" recommends including a 5-day "enrichment". The results of investigation of fresh orange juices using the unified methodology show that 5 days enrichment was required to ensure subsequent growth of all *A. acidoterrestris* strains isolated from the range of samples (Table 4-6).

4-2-4 pH

The pH profile and the optimal pH for the growth of *Alicyclobacillus* in YSG liquid medium is shown in Table 4-7 [6, 7]. The optimal pH is within the range of 3.5-4.5. YSG medium is pH 3.7 which was selected on the basis of the pH of orange juice, which is close to 3.7. The growth of *Alicyclobacillus* is affected by the type of acid used for adjusting the pH. Inhibition occurs in the following order: acetic acid >> adipinic acid > lactic acid > succinic acid. Growth was not significantly affected by fumaric acid, L-ascorbic acid, L- malic acid or citric acid [9]. For the pH adjustment of

YSG medium, sulfuric acid and citric acid showed very similar growth effects (Table 4-8, Fujita, unpublished data).

Table 4-4. Growth temperature and pH range of *Alicyclobacillus* species [17-21]

Species	Growth Temperature range (°C)	Optimum Growth Temperature (°C)	Growth pH range	Optimum pH
A. acidocaldarius subsp. *acidocaldarius*	35-70	55-60	2.5-6.0	4.5-5.0
A. acidocaldarius subsp. *rittmannii*	45-70	63	2.5-5.0	4.0
Alicyclobacillus genomic species 1	35-70	55-60	2.0-6.0	4.0-4.5
"*A. mali*"	35-65	50	2.0-5.5	4.0-5.0
A. sendaiensis	40-65	55	2.5-6.5	5.5
Alicyclobacillus genomic species 2	35-70	55-60	2.5-6.0	4.0-4.5
A. vulcanalis	35-65	55	2.0-6.0	4.0
A. acidoterrestris	20-55	40-50	3.0-6.0	3.5-4.0
Alicyclobacillus sp. 3 (= *A. fastidiosus*)	20-55	40-45	2.0-5.5	4.0-4.5
A. acidiphilus	20-55	50	2.5-5.5	3.0
A. hesperidum	35-60	50-53	2.5-5.5	3.5-4.0
Alicyclobacillus sp. 1 (= *A. saccharis*)	30-55	45-50	2.0-6.0	4.0-4.5
Alicyclobacillus sp. 2	Unknown	Unknown	Unknown	Unknown
A. cycloheptanicus	30-55	50	3.0-5.5	4.0
Alicyclobacillus sp. (= *A. macrosporangidus*)	35-60	50-55	3.0-6.5	4.0-4.5
A. pomorum	30-60	45-50	2.5-6.5	4.5-5.0
Alicyclobacillus sp. 4 (= *A. contaminans*)	35-60	50-55	3.0-6.0	4.0-4.5
Alicyclobacillus sp. 6 (= *A. kakegawaensis*)	40-60	50-55	3.0-6.5	4.0-4.5
Alicyclobacillus sp. 8 (= *A. shizuokaensis*)	35-60	45-50	3.0-6.5	4.0-4.5
A. herbarius	35-65	55-60	3.5-6.0	4.5-5.0
A. disulfidooxidans	4-40	35	0.5-6.0	1.5-2.5
A. tolerans	≦20-55	37-42	1.5-5.0	2.5-2.7

Table 4-5. Growth of three *Alicyclobacillus* species at various temperatures for 18 h

Species	Strain	30 °C	35 °C	40 °C	45 °C	50 °C	55 °C	60 °C	65 °C	70 °C
A. acidoterrestris	RB1	−	−	w+	+	+	+	−	−	−
	RB253	−	w−	w+	+	+	+	−	−	−
	RB221	−	w−	+	+	+	+	−	−	−
	RB346	−	w−	w+	+	+	+	−	−	−
	RB359	w−	w	w+	+	+	+	−	−	−
	B2065	−	w	+	+	+	+	−	−	−
	B2066	−	w	w+	+	+	+	−	−	−
	B2067	−	w	w+	+	+	+	−	−	−
Alicyclobacillus genomic species 1	DSM 11983	−	−	w−	w	+	+	+	+	w
	DSM 6481	−	−	w−	w	w+	+	+	+	w
	KHC3	−	−	w−	w	w+	+	+	+	w+
	DSM 11984	−	−	w−	w	w+	+	+	+	w
	TA3	−	−	w−	w	w	+	w+	w	w−
	DSM 453	−	−	w−	w	w	+	+	+	w
A. acidocaldarius	3B	−	−	−	w−	w−	w+	w+	w	w
	SO6	−	−	−	w−	w−	+	+	+	w
	OJ5	−	−	−	w−	w−	+	+	+	w
	P2	−	−	w−	w	w+	+	+	w	w
	HP2	−	−	w−	w	w+	+	+	+	w
	ATCC 27009^T	−	−	−	w−	w−	+	+	+	w

Strains were inoculated into Difco-based YSG agar using the plate spreading method.
+: Growth, -: no growth, w: weak growth.

Table 4-6. Results of routine microbe test of frozen concentrated orange juice

Sample	0 day		3 day		4 day		5 day	
number	30 °C	45 °C	30 °C	45 °C	30 °C	45 °C	30 °C	45 °C
1	0	54	95	194	$>10^6$	$>10^6$	$>10^6$	$>10^6$
2	0	40	0	6	0	5	0	1
3	0	210	$>10^6$	$>10^6$	$>10^6$	$>10^6$	$>10^6$	$>10^6$
4	0	690	0	175	0	180	0	55

The routine microbe test for *Alicyclobacillus* was performed using the Japanese Fruit Juice Association method. After incubation for 0, 3, 4 and 5 days at 45 °C, colony numbers were determined using the pour plating method. Colonies which grew at 30 °C were all *A. acidoterrestris*, colonies which grew at 45 °C were all *Alicyclobacillus* sp. Samples 1 and 3 are *A. acidoterrestris* positive.

Table 4-7. Differential recovery of *A. acidoterrestris* spores using two acid solutions to adjust media pH

	RB1	RB221	RB253	RB346	RB359
YSG agar (pH adjusted with H_2SO_4)	1.08×10^6	4.43×10^5	3.28×10^6	7.33×10^5	1.06×10^6
YSG agar (pH adjusted with citric acid)	9.04×10^5	6.05×10^4	1.33×10^6	3.08×10^5	4.85×10^6

YSG agar: Difco-based YSG agar

4-2-5 Heat shock conditions

Heat shock is applied in order to obtain a uniform germination of *Alicyclobacillus* spores. Eiroa and co-workers investigated the germination of *A. acidoterrestris* spores and concluded that a treatment of 70°C × 20 min was optimal for activation of spores (Table 4-9) [10]. The "Unified Method for Detection of Thermo-acidophilic bacteria" also recommends an activation period at 70°C × 20 min. Heat shock is not essential for the germination of *Alicyclobacillus* spores, perhaps due to the higher cultivation temperatures. Heat shock synchronizes germination and eliminates other microbial contaminants, such as yeasts and lactic acid bacteria. For this reason, it is not necessary to conduct a heat shock treatment when a product is tested immediately after sterilization. When a sample is tested after a long storage period under refrigerated conditions, synchronized germination is obtained after heat shock treatment.

Table 4-8. Influence of heat shock conditions on the germination of *A. acidoterrestris* spores [7]

Heat shock condition	BAM agar	OSA agar
60 °C for 30 min	66	44
60 °C for 60 min	70	50
70 °C for 20 min	90	23
80 °C for 5 min	49	26
80 °C for 10 min	78	12
80 °C for 30 min	<1	40
Boiling for 5 min	<1	<1

Colony numbers were measured after incubation at 50 °C for 6 days.

4-2-6 Effect of concentration and the type of juice

The type of juice greatly affects the growth of *Alicyclobacillus*. Splittstoesser and co-workers have investigated the growth of two strains of *A. acidoterrestris*-like bacteria in different types of fruit juices [11]. The inoculated strains grew very well in apple juice, tomato juice and white grape juice. Opposite results were observed with red grape juice, in which very strong growth inhibition was observed (Table 4-10). Goto and co-workers tested the growth behavior of several strains of *A. acidoterrestris* in a variety of fruit juices, and have concluded that the behavior of the strains depends on the type of juice and also on the source of isolation of the strains (Table 4-11) [12].

The growth of *Alicyclobacillus* is also affected by the juice concentration. Members of the JFJA Working Group investigated the relationship between the juice concentration and growth characteristics by examining the growth profile of the bacteria in several fruit juices (Table 4-12) [12].

The results showed that there was a strong correlation with the type of juice, however even with orange juice, which showed good growth promotion, there was a tendency toward complete inhibition of growth when the juice concentration was over 50-60 % [13]. For juices with a high polyphenol content, such as red grape juice, detection of *Alicyclobacillus* was difficult even when the juice concentration was over 10-20 % [14]. Based on these results, the dilution ratio recommended in the unified methodology for detection of the thermo-acidophilic bacteria was determined (Table 4-12) [12].

Table 4-9. Growth of *A. acidoterrestris* in commercial beverages [11]

Type of beverage	pH	Brix	Growth	
			Strain VF	Strain WAC
Apple juice	3.5	11-11.4	+	+
Tomato juice	4.0	7.0	+	+
Apple-grape juice	2.8-3.7	12.2-14.8	-	-
Apple-orange-pineapple juice	2.9	14.8	-	+
Apple-grape-cherry juice	3.7	12.4	-	-
Red grape juice	2.9-3.3	13.6-15.8	-	-
Cranberry juice	2.4	14.0	-	-
Grapefruit juice	3.2	10.4	+	-
Orange juice	3.6	12.0	+	-
Riesling (white grape) juice	3.4	10.8	+	+
Elvira (white grape) juice	3.4	7.8	+	+
Cabernet sauvignon (red grape) juice	3.7	12.2	-	-
Pinotage (red grape) juice	3.8	10.9	-	-
Pineapple juice	3.3	13.4	+	-
Prune juice	3.7	18.8	-	-

Table 4-10. Growth of *A. acidoterrestris* in beverages [9]

	A. acidoterrestris strain								
	B2065	B2066	B2067	RB1	RB221	RB253	RB346	RB359	TAB H1
30 % apple juice	+	+	+	-	+	-	+	+	+
30 % apple juice	+	+	-	+	-	+	+	+	+
30 % lemon juice	-	+	-	+	+	-	+	+	-
30 % grape fruit juice	-	+	+	+	+	-	+	+	+
100 % mix fruit juice	+	+	+	+	+	+	+	+	+

B2065-B2067: isolated from apple juice, RB1, RB253 and RB346: isolated from orange juice, RB221: isolated from isotonic drink, RB359: isolated from mango juice, TAB H1: isolated from pineapple juice. +: Growth, -: no growth.

Table 4-11. Upper limit of juice concentration (%) for growth of *Alicyclobacillus* [9]

	A. acidoterrestris		*A. acidocaldarius*		*Alicyclobacillus* genomic species 1	
Pre-incubation temperature	45 °C	55 °C	45 °C	55 °C	45 °C	55 °C
Orange juice	≦30～50	-	-	≦30	≦20	≦50
Grapefruit juice	≦30	-	-	≦20	≦10	≦30～50
Apple juice	≦50	-	-	≦30	≦20	≦30
Peach juice	≦50	-	-	≦10	≦30	≦50
Red grape juice	10	n.d.	n.d.	n.d.	n.d.	n.d.
Prune juice	20	n.d.	n.d.	n.d.	n.d.	n.d.
Blueberry juice	10	n.d.	n.d.	n.d.	n.d.	n.d.

Test was performed based on the Japanese Fruit Juice Association method. -: no growth, n.d.: no data.

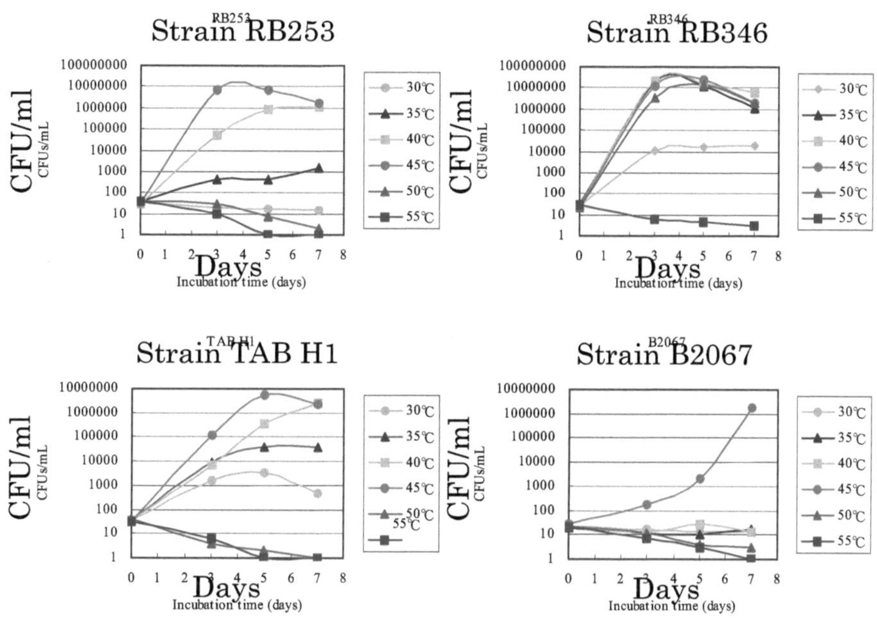

Fig. 4-4. Growth of four *A. acidoterrestris* strains in 100 % orange juice

A. acidoterrestris spores (20-50 CFU/ml) were inoculated into 200 ml of orange juice and then incubated at 30-55 °C for 7 days.

Table 4-12. Influence of heat shock conditions on the germination of *A. acidoterrestris* spores [7)]

Fruit juice	Dilutio ratio
Apple juice (Brix 70 %)	10 times
Orange juice (Brix 65 %)	10 times
Red grape juice (Brix 65 %)	20-50 times
White grape juice (Brix 65 %)	20 times
Carrot juice (Brix 40 %)	10 times
Lemon juice (Brix 50 %)	10 times
Grapefruit juice (Brix 55 %)	20 times
Peach juice (Brix 55 %)	10 times

4-3 Unified method for detection of thermo-acidophilic bacteria

A unified methodology for the detection of thermo-acidophilic bacteria has been jointly designed by a consortium of ten beverage companies and the JFJA. This method uses YSG medium (pH 3.7±1), with an incubation temperature close to the common growth temperature for *Alicyclobacillus* (45°C), improving the selective growth of *Alicyclobacillus*. For samples that can be filtered, "Membrane Filtration Method" can be applied (4-3-1). "Pour Plating Method" or "Spread Plating Method" can be used for samples that are not suitable for "Membrane Filtration Method" (4-3-2). Samples that are suitable for "Membrane Filtration Method" can also be tested by "Pour Plating Method" or "Spread Plating Method". Differentiation of detected colonies can be verified by Peroxidase Kit (Va-YSG Medium, Code 08900, Kyokuto Pharmaceutical Co., Ltd., Japan) and the Guaiacol Detection Kit (Code 08920, Kyokuto Pharmaceutical Co., Ltd., Japan) (4-3-3). Alternatively, differentiation can be carried out using the "Differential Temperature Test" (4-3-3). Target organisms which produce guaiacol can be detected using the Peroxidase Kit (4-3-4).

Figures 4-5, 4-6 and 4-7 explain in detail the test method steps for *Alicyclobacillus*. Basically, an analyte (usually concentrated fruit juice) is diluted in sterile water for the "Membrane Filtration Method" or YSG liquid medium for other methods. The dilution varies among juice type, as shown in Table 4-12. The sample is heat shocked at 70°C for 20 min, cooled immediately and then cultured using one of the steps below (4-3-1, 4-3-2 or 4-3-4).

4-3-1 Clarified juice or syrup (liquid sugar)

Filter through a 0.45 μm membrane, transfer the membrane onto a YSG agar surface, incubate at 45°C for 3-5 days, and then count the colonies formed (Fig 4-5).

4-3-2 Turbid fruit juice

Pre-incubate the sample at 45°C for 3-5 days (enrichment), and then determine cfu using the "Pour Plating Method" or the "Spread Plating Method" (Fig 4-6). To determine the initial number of cells, take an aliquot of the sample prior to enrichment and culture it using the "Pour Plating Method" or the "Spread Plating Method".

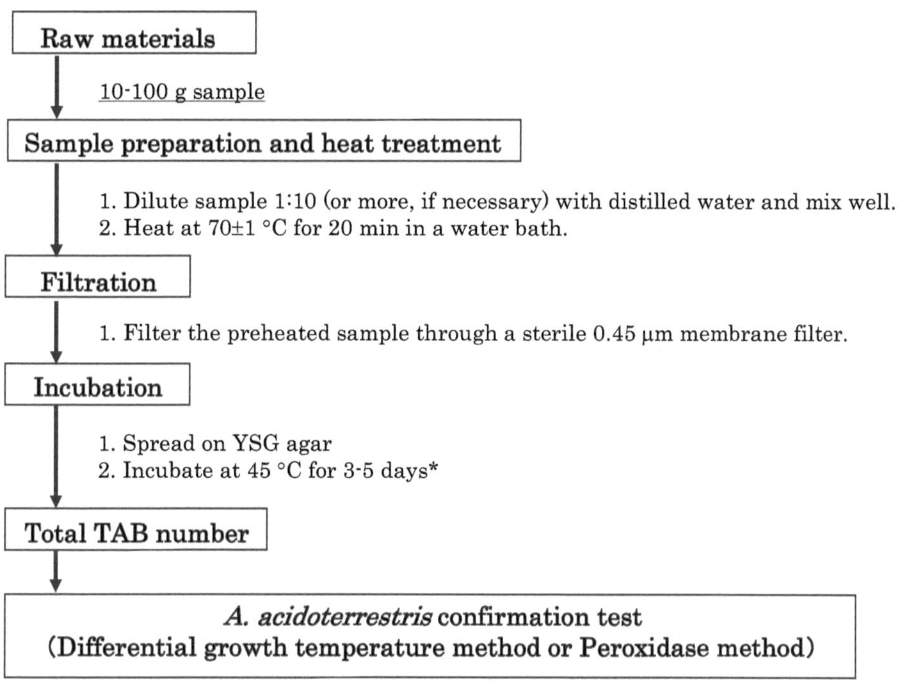

Fig. 4-5. Standardized Membrane method for detection of thermo-acidophilic bacilli

*1: Recommend 5 days incubation.

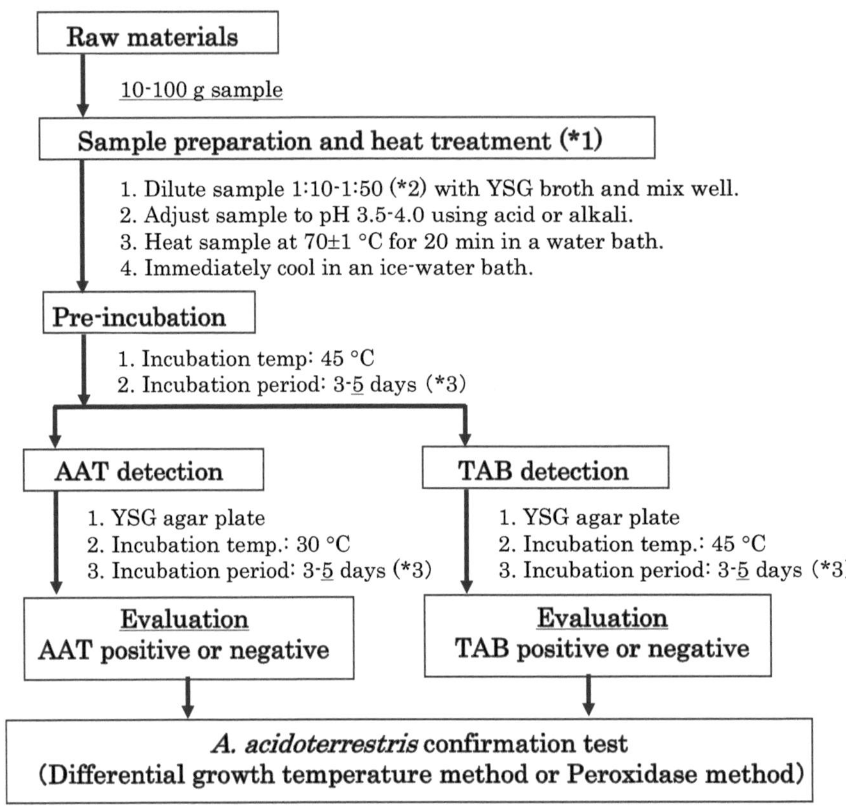

Fig. 4-6. Standardized Pour plating and Spreading methods for the detection of Thermo-Acidophilic Bacilli

In order to investigate initial numbers, test small amount of sample before pre-incubation.
*1: Keep headspace, *2: see Table 4-14, *3: recommend 5 days incubation.

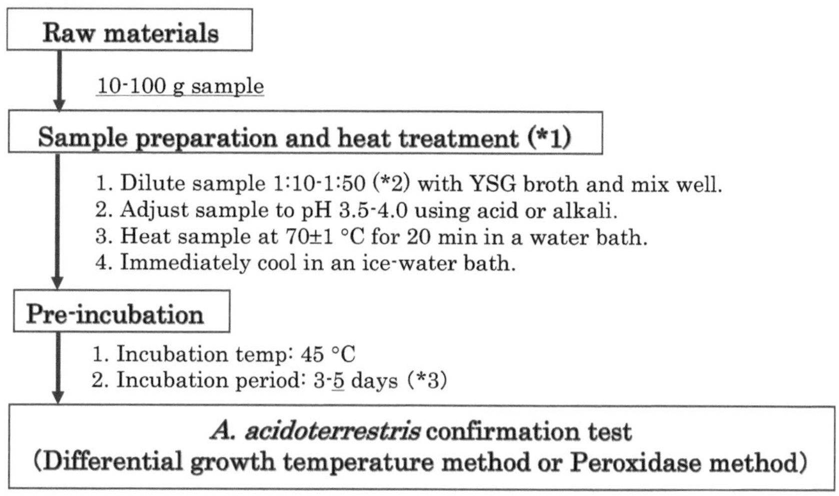

Fig. 4-7. Standardized Direct detection method for thermo-acidophilic bacilli

In order to investigate initial numbers, test small amount of sample before pre-incubation.
*1: Keep headspace, *2: see Table 4-14, *3: recommend 5 days incubation.

4-3-3 Differentiation of species

When the detection of guaiacol-producing species is required, fish colonies with a sterile loop and perform the "Peroxidase Test" or "Differential Temperature Method" (See "Chapter 5, Identification and differentiation methods of *Alicyclobacillus*").

4-3-4 Simplified method

After incubating the sample at 45°C for 5 days (enrichment), test the sample with the "Peroxidase Test" or the "Differential Temperature Method", and determine the presence or absence of guaiacol producers (Fig 4-7 and Fig 4-8). To determine the initial number of cells, dilute an aliquot of the sample and culture it using the "Pour Plating Method" or the "Surface Plating Method".

Detailed protocols for these methods are described in the Bulletin of the Fruit Juice Association in 2003 March (in Japanese) [2], 2003 April (in English) [15].

1. Dilute sample with YSG broth, for dilution ratio see Table 4-14.
2. Adjust sample to pH 3.5-4.0 using acid or alkali.
3. Heat sample at 70±1 °C for 20 min in a water bath.
4. Immediately cool in an ice-water bath.
5. Incubate at 45 °C for 3-5 days.
6. Check AAT using the following method.

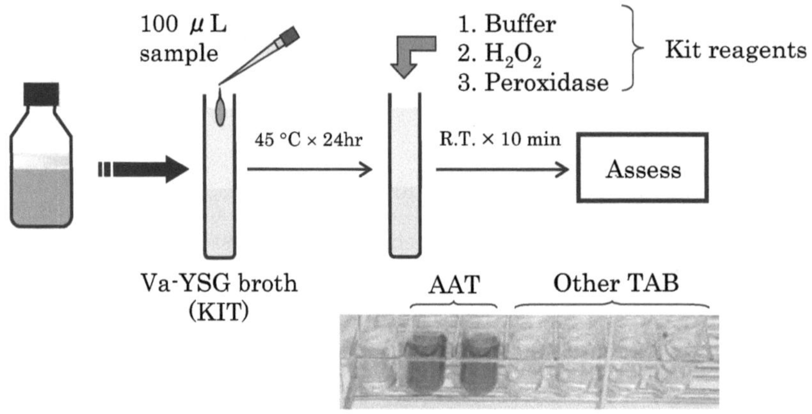

Fig. 4-7. Standardized thermo-acidophilic bacilli direct detection method using a peroxidase kit

(Addendum 1) YSG liquid medium

(1) Dissolve yeast extract (2 g), glucose (1 g) and soluble starch (2 g) in 1000 mL of water, then adjust the pH to 3.7±0.1 with sulfuric acid or hydrochloric acid.
(2) Aliquot the medium in bottles, then autoclave (121°C × 15 min).

(Addendum 2) YSG agar medium

(1) Prepare solutions A and B and autoclave separately (121°C × 15 min).
Solution A: Dissolve yeast extract (2 g), glucose (1 g) and soluble starch (2 g) are dissolved in 500 mL of water, then adjust the pH to 3.7±0.1 with sulfuric acid or hydrochloric acid.
Solution B: Dissolve 15 g of agar in 500 mL water.
(2) Allow solutions A and B to cool to 50-60°C before combining. Dispense medium into Petri dishes, solidify then allow the surface to dry.
*For "Pour Plating Method", keep the medium at 50°C after mixing the solutions.

4-4 IFU Method on the detection of taint producing *Alicyclobacillus* in fruit juices

"The IFU Method on the Detection of taint producing *Alicyclobacillus* in Fruit Juices, revision 1", September 2004 was developed by the IFU Working Group of Microbiology, and is based on the use of K-agar in combination with BAT medium (pH 4.0) or YSG. Samples that are not easily membrane filtered are inoculated by "Spread Plating Method" [>100 CFU/g (4-4-1)], or by performing a pre-enrichment [samples with <100 CFU/g (4-4-2)]. Analytes that can be filtered should be analyzed by the "Membrane Filtration Method" (4-4-3). Methodologies to analyze industrial products are also included [see (4-4-4) and (4-4-5)]. Differentiation is based on the inability of *Alicyclobacillus* to grow on standard agar at neutral pH, growth and spore production at 45°C on media at pH <4.0 (K-agar and YSG/BAT) and growth at 65°C on YSG. Taint-producing strains of *Alicyclobacillus* are unlikely to grow at 65°C (4-4-6).

Method flow diagrams are shown in Figs 4-9, 4-10, 4-11, 4-12 and 4-13. The outline of the test is as follows: the analyte is diluted 10-fold with an appropriate diluent, heat shock treated at 80°C for 10 min and then immediately cooled to 40-45°C This pre-treatment flow is common to the other samples. According to the species of the bacteria or the expected microbial count, the test method can be selected from 4-4-1, 4-4-2 or 4-4-3. Finished processed products are assayed using conditions given in 4-4-4 or 4-4-5.

For further details see revision 1 (September 2004) of the IFU method.

4-4-1 Juice concentrates and other raw materials that can not be easily membrane filtered (>100 CFU/g)

0.1 mL of sample is spread on K-agar and on YSG or BAT agar, and then incubated for 2-5 days at 45±1°C (Fig. 4-9).

4-4-2 Juice concentrates and other raw materials that can not be easily membrane filtered (<100 CFU/g)

Incubate the sample at 45±1°C for 5 days (enrichment). After 2 days of enrichment an aliquot of 0.1 mL is spread plated on K-agar and YSG or BAT agar and incubated for 2 days at 45±1°C. If there is no growth on these plates, repeat after 5 days of enrichment using fresh plates (Fig. 4-10).

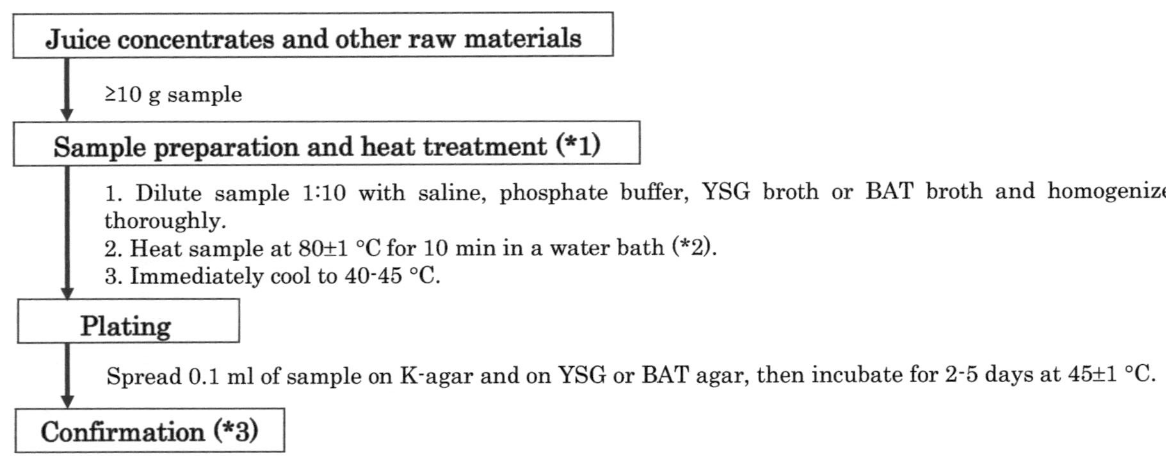

Fig. 4-9. IFU method for the detection of taint-producing *Alicyclobacillus* in non-filterable samples (> 100 cfu/ml)

*1: For juice concentrates, sterile distilled/demineralized water is an acceptable diluent and provides an environment similar to that of a final juice product. For sugar solutions or water etc., additional nitrogen or carbon sources may be required to support bacterial growth. BAT or YSG broths may provide these nutrients, but will also provide additional minerals that are not generally present in most juice products.
*2: The heat shock treatment is intended to activate bacterial spores and inactivate vegetative cells.
*3: See text.

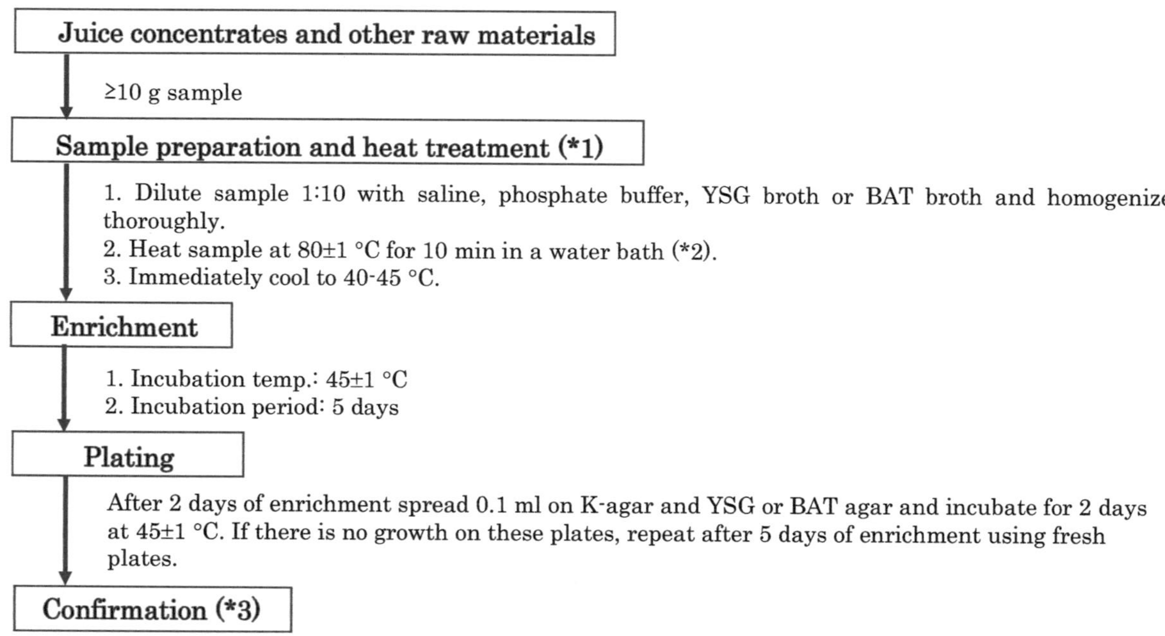

Fig. 4-10. IFU method for the detection of taint producing *Alicyclobacillus* in non-filterable samples (< 100 cfu/ml)

*1: For testing juice concentrates sterile distilled/demineralized water is an acceptable diluent and provides an environment similar to that of a final juice product. For sugar solutions and water etc additional nitrogen or carbon sources may be required to support bacterial growth. BAT or YSG broths may provide these nutrients, but also provide additional minerals that are not generally present in most juice products.
*2: The heat shock treatment is intended to activate bacterial spores and inactivate vegetative cells. *3: See text.

4-4-3 Juice concentrates and other raw materials that can be membrane filtered

Filter sub-samples of the sample through a 0.45 μm pore size membrane. Lay the filter on the surface of K-agar and YSG or BAT agar, and incubate at 45±1°C for 2-5 days (Fig. 4-11).

4-4-4 Final products directly after processing

Incubate the product in its original container at 45°C for 7 days, then spread plate a sample on K-agar and YSG or BAT agar surface and incubate at 45±1°C for 2-5 days (Fig. 4-12).

4-4-5 Final products taken from the market

Inoculate 0.1 mL of the product onto K-agar and YSG or BAT medium agar, and incubate at 45±1°C for 2-5 days (Fig. 4-13). If necessary, pre-enrich samples at 45±1°C for 7 days.

4-4-6 Confirmation tests

See revision 1 (September 2004) of the IFU method:

- Examine colonies growing on K agar, YSG agar and BAT agar. Select a representative number of each colony type on which to conduct confirmatory tests.
Note: YSG and BAT medium support the growth of all known species of *Alicyclobacillus*. It is therefore likely that a broader range of colony types will be visible on these media compared to K agar.

- From each selected colony, streak a portion onto one K agar plate, one plate containing neutral pH medium (e.g. plate count agar, tryptic soy agar or brain heart infusion agar) and two YSG agar plates. Incubate the plate containing neutral pH medium, K agar and one YSG plate at 45±1°C for 3-5 days. Incubate the second YSG plate at 65±1°C for 2-3 days.

- Inspect plates for growth. There should be no growth on the neutral pH medium. If growth is observed on this medium, record the result as negative for *Alicyclobacillus*.

- Examine colonies on K agar for the presence of spores by examining a wet mount with phase contrast microscopy, or perform a spore stain. If there is no growth on K agar, growth should be observed on YSG agar. Spore formation may not occur on YSG agar, so spore formation on an additional medium such as BAT agar may be required, preferably at pH <4.0 to exclude other acid-tolerant spore-forming bacteria such as *Bacillus coagulans*.

- Examine the YSG plate incubated at 65°C. Taint producing strains of *Alicyclobacillus* are unlikely to grow at 65°C. If growth is observed, record the result as negative for presumptive taint producing *Alicyclobacillus.*

- Record colonies that produce spores at pH 3.7 and do not grow at pH >6.0 as presumptive *Alicyclobacillus*. Those that do not grow at 65°C are presumptive taint producing *Alicyclobacillus*.

(Addendum) BAT agar

(1) Prepare solutions A and B, and autoclave separately (121°C × 15min).
Solution A: $CaCl_2 \cdot 2H_2O$ (0.25 g), $MgSO_4 \cdot 7H_2O$ (0.50 g), $(NH_4)_2SO_4$ (0.20 g), KH_2PO_4 (3.0 g), yeast extract (2.0 g), glucose (5.0 g), Trace Minerals Solution* 1 ml are dissolved in 500 mL water, and adjusted to pH 4.0 with 1N H_2SO_4 or 1N NaOH.
*Trace Minerals Solution for Solution A: $CaCl_2 \cdot 2H_2O$ (0.66 g), $ZnSO_4 \cdot 7H_2O$ (0.18 g), $CuSO_4 \cdot 5H_2O$ (0.16 g), $MnSO_4 \cdot H_2O$ (0.15 g), $CoCl_2 \cdot 5H_2O$ (0.18 g), H_3BO_3 (0.10 g), $Na_2MoO_4 \cdot 2H_2O$ (0.30 g) are dissolved in 1 L water then sterilized at 121°C for 15 min. This solution can be stored at 4°C.
Solution B: Dissolve agar (15-20 g) in 500 mL water. Sterilize by autoclaving at 121°C for 15 min.
(3) Allows solutions A and B to cool to 50°C before combining. Dispense medium into petri dishes, solidify then allow the surface to dry.

Juice concentrates and other raw materials

↓ ≥10 g sample

Sample preparation and heat treatment (*1)

1. Dilute sample 1:10 with saline, phosphate buffer, YSG broth or BAT broth and homogenize thoroughly.
2. Heat at 80±1 °C for 10 min in a water bath (*2).
3. Immediately cool to 40-45 °C.

↓

Filtration

*If possible, filter the diluted and heated sample through a 0.45 μm pore size membrane. This allows colonies to be counted and provides a detection limit of 1 cfu/100 ml diluted sample or 1 cfu/10 g of the original sample. Prepare 2×100 ml diluted and heated samples so that two solutions can be filtered and incubated on two membranes. Filter each sample aseptically and wash the sides of the filter apparatus with around 20 ml rinse solution (such as sterile distilled/demineralized water). Transfer the first membrane aseptically onto K agar and transfer the second membrane onto one of YSG or BAT agars. Incubate at 45±1 °C for 2-5 days.

↓

Confirmation (*3)

Fig. 4-11. IFU method for the detection of taint producing *Alicyclobacillus* in filterable samples

*1: For testing juice concentrates sterile distilled/demineralized water is an acceptable diluent and provides an environment similar to that of a final juice product. For sugar solutions and water etc additional nitrogen or carbon sources may be required to support bacterial growth. BAT or YSG broths may provide these nutrients, but also provide additional minerals that are not generally present in most juice products.
*2: The heat shock treatment is intended to activate bacterial spores and inactivate vegetative cells. *3: See text.

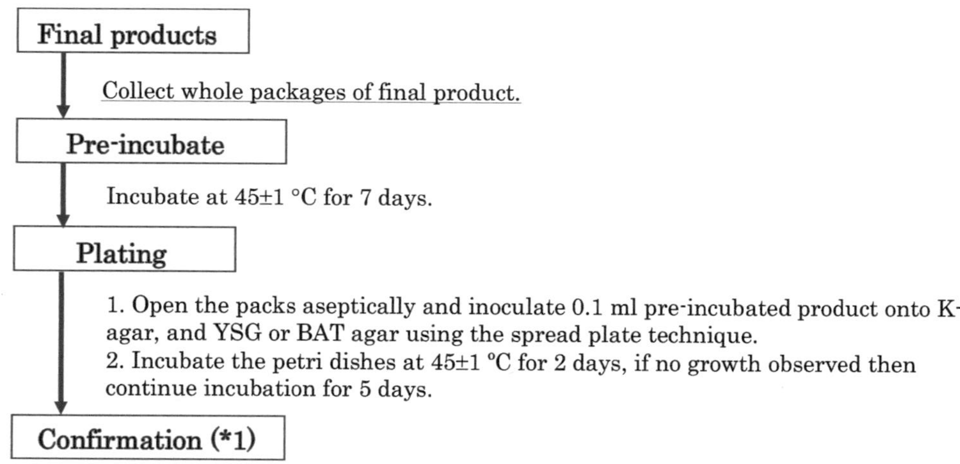

Fig. 4-12. IFU method for the detection of taint producing *Alicyclobacillus* in final products sampled directly after processing

*1: See text.

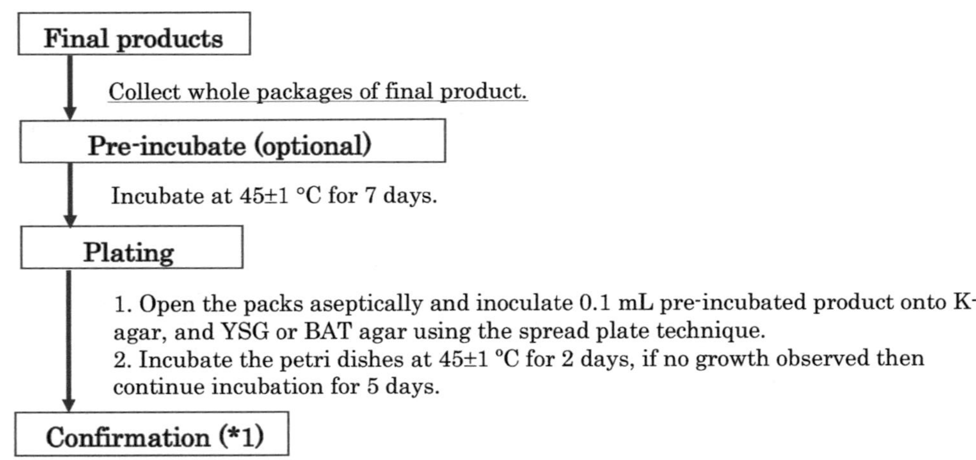

Fig. 4-13. IFU method on the detection of taint producing *Alicyclobacillus* in final products taken from the market

*1: See text.

4-5 Guidelines for *Alicyclobacillus*

Guidelines relating to *Alicyclobacillus* detection are not defined internationally or within Japan. *A. acidoterrestris* is considered the most important species for quality control purposes in Japan. However, other members of the *Alicyclobacillus* genus can produce guaiacol and cause deterioration of manufactured foodstuffs. Accordingly, methods for the detection of all species need to be defined for routine use. Spoilage may occur from a single spore in a 500 mL bottle [14], meaning detection methods must be extremely sensitive. The characteristics of growth varies among juice types and according to the species and strain (Table 4-15). The IFU has determined taint- producing *Alicyclobacillus* to be the main target organism.

Keiichi Goto

References

1. Cerny G, Hennlich W, Poralla K (1984) Fruchtsaftverderb durch Bacillen: Isolierung und Charakterisierung des Verderbserregers. Z Lebens Unters. Forsch 179:224-227 (In German).
2. Nippon Kaju Kyoukai (2003) "Tainetsu sei kousanseikin touitsu kensahou" ni tsuite. Kaju Kyoukai Hou 3:4-12 (in Japanese).
3. Walls I, Chuyate R (2000) Spoilage of fruit juices by *Alicyclobacillus acidoterrestris*. Food Australia 52:286-288.
4. Walls I, Chuyate R (2000) Isolation of *Alicyclobacillus acidoterrestris* from fruit juices. J AOAC Int, 83:1115-1120.
5. Goncalves AC (2003) ILSI Japan, Symposium of Thermo-acidophillic bacteria, "Comprehension of TAB Research of ABECitrus and TAB control now and in the future".
6. Tsuruoka N, Isono Y, Shida O, Hemmi H, Nakayama T, Nishino T (2003) *Alicyclobacillus sendaiensis* sp. nov., a novel acidophilic, slightly thermophilic species isolated from soil in Sendai, Japan. Int J Syst Evol Microbiol 53:1081-1084.
7. Goto K, Mochida K, Asahara M, Suzuki M, Kasai H, Yokota A (2003) *Alicyclobacillus pomorum* sp. nov., a novel thermo-acidophilic, endospore-forming bacterium that does not possess ω-alicyclic fatty acids, and emended description of the genus *Alicyclobacillus*. Int J Syst Evol Microbiol 53:1537-1544.
8. Goto K (2003) ILSI Japan, Symposium of Thermo-acidophillic bacteria, "Saishin tainetsusei kousansei kin no kisochishiki to sono taisakuhou" (in Japanese).

9. Yamashita M (1996) Arukoru rui no biseibutsu ni taisuru sayou. Boukin Boubai 24:25-49 (in Japanese).
10. Eiroa MNU, Junqueira VCA, Schmidt F (1999) *Alicyclobacillus* in orange juice: occurrence and heat resistance of spores. J Food Protection 62:883-886.
11. Splittstoesser DF, Lee CY, Churey JJ (1998) Control of *Alicyclobacillus* in the juice industry. Dairy Food Environ. Sanitation 18:585-587.
12. Goto K (2003) Tainetsusei kousanseikin no kensahou - Bunseki hou touitsu to hatten no doukou <Tainetsusei kousanseikin touitsu kensahou no kensaseido oyobi Q&A>. Kaju Kyoukai Hou 10:1-16. (in Japanese).
13. Niwa M (2004) Guaiacol kenshutsu kit (Peroxidase hou) ni yoru sansei inryou yuugakin no kanri. Japan Food Science 43:23-28 (in Japanese).
14. Mitsuda H (2003) Tainetsusei kousanseikin no kensahou - Bunseki hou touitsu to hatten no doukou <Kensahou touitsu he no ugoki>. Kaju Kyokai Hou 11:1-8 (in Japanese).
15. Nippon Kaju Kyokai (2003) The unified detection method of thermo-acidophilic bacteria (TAB) in raw materials of fruit juice. Kaju Kyoukai Hou 4:19-27.
16. IFU Method on the Detection of taint producing *Alicyclobacillus* in Fruit Juices, revision 1, September 2004 (IFU Handbook Microbiological Methods, Method No.12).
17. Tsuruoka N, Isono Y, Shida O, Hemmi H, Nakayama T, Nishino T (2003) *Alicyclobacillus sendaiensis* sp. nov., a novel acidophilic, slightly thermophilic species isolated from soil in Sendai, Japan. Int J Syst Environ Microbiol 53:1081-1084.
18. Goto K, Mochida K, Asahara M, Suzuki M, Kasai H, Yokota A (2003) *Alicyclobacillus pomorum* sp. nov., a novel thermo-acidophilic, endospore-forming bacterium that does not possess ω-alicyclic fatty acids, and emended description of the genus *Alicyclobacillus*. Int J Syst Evol Microbiol 53:1537-1544.
19. Goto K, Mochida K, Kato Y, Asahara M, Fujita R, An SY, Kasai K, Yokota A Proposal of *Alicyclobacillus contaminans* sp. nov., *Alicyclobacillus fastidiosus* sp. nov., *Alicyclobacillus kakegawaensis* sp. nov., *Alicyclobacillus macrosporangidus* sp. nov., *Alicyclobacillus sacchari* sp. nov., and *Alicyclobacillus shizuokaensis* sp. nov., novel moderately-thermophilic, acidophilic, endospore-forming bacteria. Int J Syst Evol Microbiol (in review).
20. Simbahan J, Drijber R, Blum P (2004) *Alicyclobacillus vulcanalis* sp. nov., a thermophilic, acidophilic bacterium isolated from Coso Hot Spring, California, USA. Int J Syst Evol Microbiol 54:1703-1707.
21. Karavaiko GI, Bogdanova TI, Tourova TP, Kondrat'eva TF, Tsaplina IA, Egorova MA, Krasil'nikova EN, Zakharchuk LM (2005) Reclassification of '*Sulfobacillus thermosulfidooxidans* subsp. *thermotolerans*' strain K1 as *Alicyclobacillus tolerans* sp. nov. and *Sulfobacillus disulfidooxidans* Dufresne et al. 1996 as *Alicyclobacillus disulfidooxidans* comb. nov., and emended description of the genus *Alicyclobacillus*. Int J Syst Evol Microbiol 55:941-947.

Chapter 5: Differentiation and identification of *Alicyclobacillus* species

5-1 Introduction

Over the past few years, various methods have been developed for the differentiation and identification of *Alicyclobacillus*. These methods are described in sections 5-2 to 5-13. The details of each method are provided however the methods differ slightly from each other in sensitivity and performance. Therefore it is very important to have a clear idea of the desired objective before selecting the analytical method to be applied. For example, the peroxidase method is effective for the detection of guaiacol and the differentiation of guaiacol producing species (5-2). The 16S rRNA gene (16S rDNA) sequence comparison method is effective for species identification (5-4), and ribotyping is the preferred method for identification at the strain level (5-7).

Prompt evaluation of the samples and measures to avoid cross contamination are essential in order to perform proper differentiation and identification of the species. When the vegetative cells are healthy, it is easier to isolate the target bacteria. However, when a long period of time (more than 10 days at 4°C or 45°C) has elapsed after active growth, the cell count will be extremely low (or cells may become completely inactive or dead). In such cases, isolation of viable cells will be very difficult or impossible. Also, when other microorganisms are present (such as from secondary contamination after opening the container), the process of isolation becomes more difficult. In these cases, if the spores are still viable, a heat shock step can be used for selective recovery. If the spores are dead, the best way to evaluate the sample is to detect the off flavor components (guaiacol and other phenolic components) in the spoiled product by using the peroxidase method or gas chromatography.

5-2 Peroxidase method

The Peroxidase method, which was developed by Niwa and co-workers in Japan (Kirin Beverage Corp.) in 2002, is based on the detection of guaiacol produced by *A. acidoterrestris* and the other guaiacol producers, by means of an enzymatic methodology [1]. This is very useful for differentiation of colonies on an agar plate. It is also very handy to differentiate guaiacol producing species when used in combination with other routine tests such as the unified method for the detection of TAB (see 4-3 the Unified Method of Japan Fruit Juice Association). This method can also be used for the analysis of other non-*Alicyclobacillus* guaiacol producers such as *Bacillus subtilis* [2].

The principle of the method is as follows: guaiacol is produced through the metabolism of vanillic acid (precursor of guaiacol) by *A. acidoterrestris* and the other guaiacol producers in the detection medium. Four molecules of guaiacol are converted into tetraguaiacol in the presence of H_2O_2 and peroxidase (Fig. 5-1). Measurement of the OD_{470nm} allows for the quantitative determination of tetraguaiacol (when values are over 0.1 in relation to the control, or by observing the difference in the browning of the suspension). A diagram of the analytical method is shown in Fig. 5-2. To obtain conclusive results, it is best to use several fresh colonies for the inoculation. Recently, this test has been commercialized in the form of an analysis kit by Kyokuto Pharmaceutical Co., Ltd. (Japan). Use of this kit can reduce the work necessary to adjust and validate the medium and other reagents.

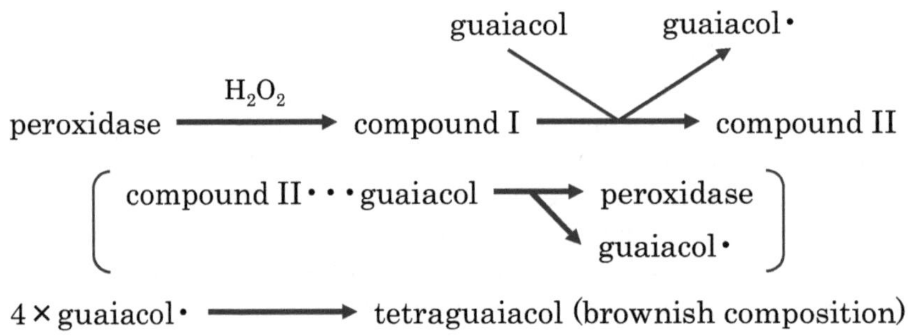

Fig. 5-1. Principal of peroxidase method

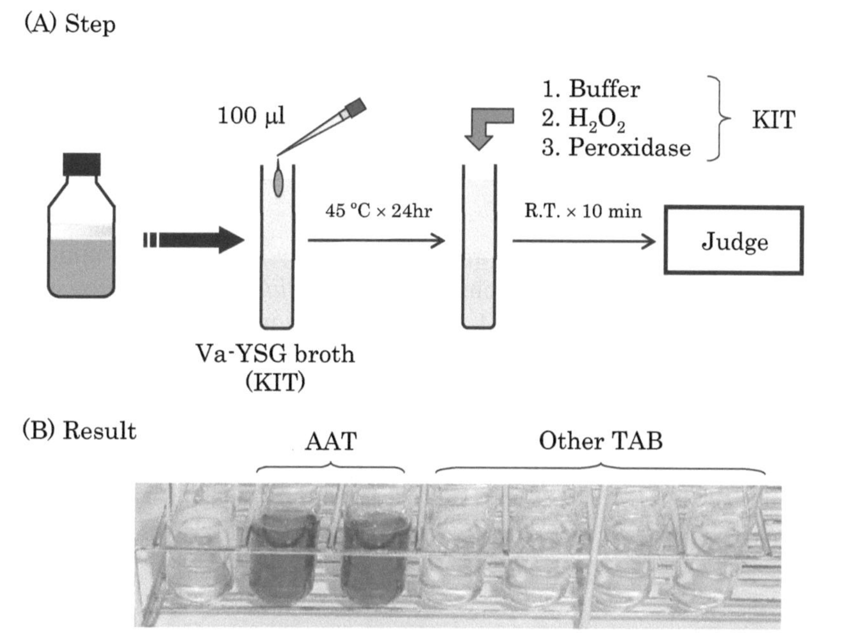

Fig. 5-2. Detection of guaiacol producing species using colony by peroxidase method

5-3 Method on the difference of growth temperature

This method is based on the fact that *A. acidoterrestris* has a relatively low optimum temperature for growth, when compared to the other *Alicyclobacillus* species. This method allows for differentiation among the *A. acidocaldarius* group and the *A. acidoterrestris* group (including *A. acidiphilus* and *A. hesperidum*), within a period of 24 hours.

In practice, the method consists of smearing a loopful of a colony, or a mixture of colonies onto two YSG agar plates, incubating one plate at 45°C and the other at 65°C and observing their growth after 18-20 h (see Fig. 5-3). The interpretation of the results would be as follows (see Fig. 5-4):

1. Clear colony growth at 45°C, and no observation of growth at 65°C: positive for *A. acidoterrestris*.

2. No colony growth at 45°C, and observation of growth at 65°C: negative for *A. acidoterrestris* and positive for the *A. acidocaldarius* group.

3. Clear colony growth at both 45°C and 65°C: positive for *A. acidoterrestris* and also positive for the *A. acidocaldarius* group (this means both species are present in the sample).

4. No colony growth observed at either temperature: negative for *A. acidoterrestris* and positive for *A. acidocaldarius* group (mainly *Alicyclobacillus* genomic species 1).

The group comprised of *A. acidiphilus*, *A. cycloheptanicus* and *A. hesperidum* exhibits the same behavior as *A. acidoterrestris*. However, *A. cycloheptanicus* and a part of *A. hesperidum* do not produce guaiacol. The main guaiacol producers are included, so we may conclude that from a quality standpoint this would not affect the result (since the isolation of *A. herbarius* is very rare, this strain was excluded from the present list of quality control indicators).

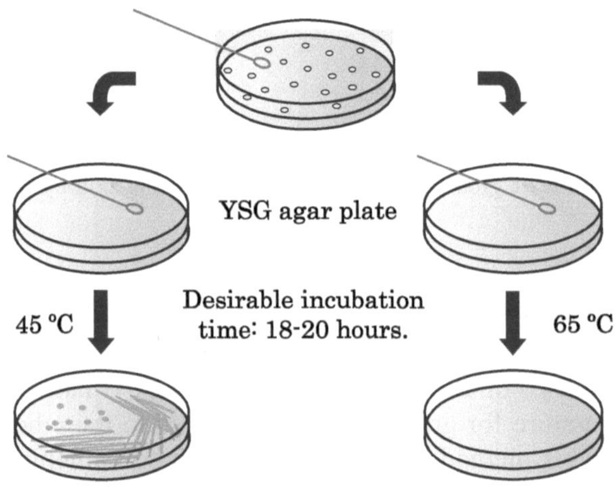

Fig. 5-3. Step of differential temperature method

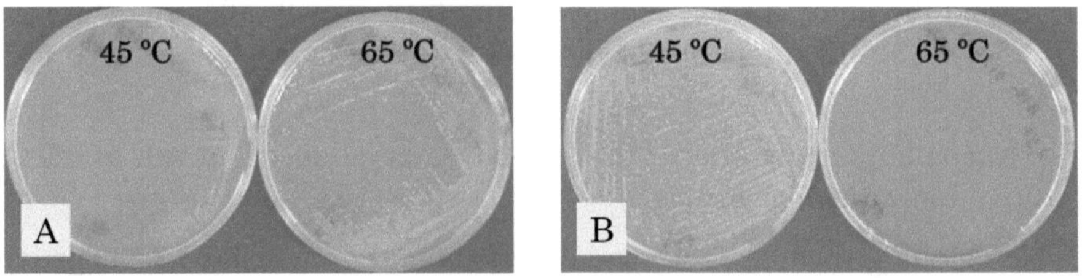

Fig. 5-4. Colony formation pattern in differential temperature method

5-4 Method of 16S rDNA sequencing

This method is based on the differences present in the nucleotide sequence of 16S rDNA, especially the hypervariable (HV) region of 16S rDNA (about 300 bp), among *Alicyclobacillus* species. It is important to work with an isolated and pure colony. Presently this is the most accurate and reliable technique for identification at the species level [3-5].

From a practical point of view, (limited to the HV region analysis), the isolated pure culture is cultivated on YSG medium for 24 h, and DNA is extracted from one loopful of a colony using commercially available kits. Using a specific primer for the HV region amplification, the fragment is amplified by PCR. The sequencing reaction is performed using the amplified fragment and the nucleotide sequence of the HV region is determined. Finally, a search of similar sequences is performed using the Blast System of the National Genome Research Gene Bank (DDBJ: http://www.ddbj.nig.ac.jp/), National Center for Biotechnology Information (NCBI: http://www.ncbi.nlm.nih.gov/) and European Bioinformatics Institute (EMBL: http://www.ebi.ac.uk/embl/), and based on these results the most similar species is chosen.

Applied Biosystems (USA) has recently been commercializing a complete kit for this analysis and database. The time required for this analysis for several strains is at least 12 h (depending on the number of strains). The phylogenetic tree constructed from the results obtained using this tool is shown in Fig. 5-5. By constructing phylogenetic trees like this, it is possible to define similarity proximity, allocating each species to a position in relation to the other species, even new species. The positive aspect of this method is its applicability for the identification of *Alicyclobacillus* as well for the other bacteria [6-8]. The general scheme for this method is presented in Fig. 5-6.

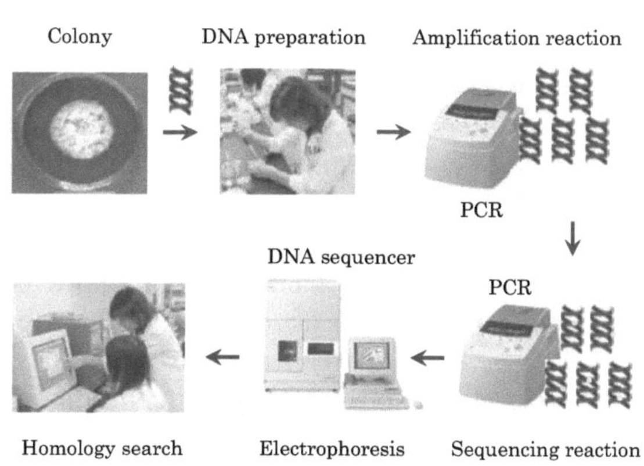

Fig. 5-6. Step of 16S rDNA analysis method

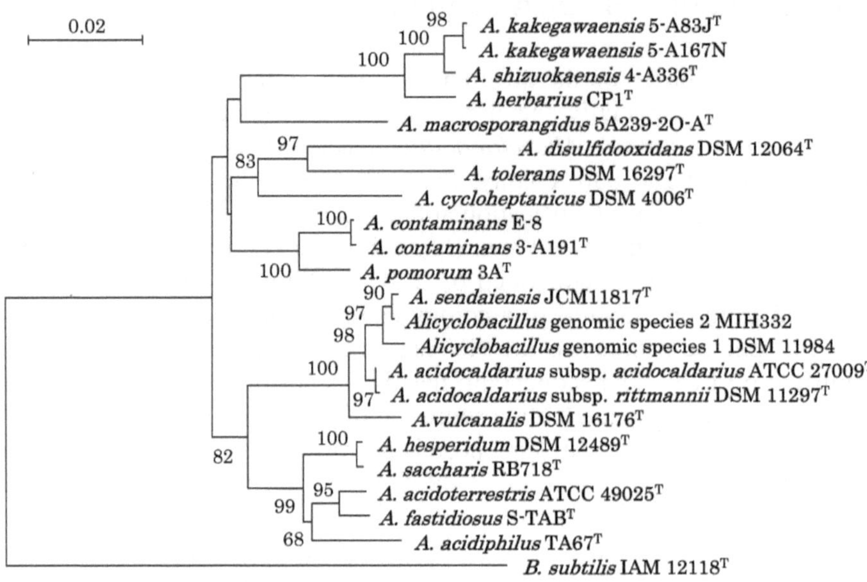

Fig. 5-5. Phylogenetic tree (constructed using the maximum parsimony method) of *Alicyclobacillus* species and related species based on 16S rDNA

Bacillus subtilis IAM 12118^T was served as outgroup. The final dataset included 1468 unambiguously aligned sites. Numbers represent percentages from 1000 replicate bootstrap samplings (frequencies of less than 70 % are not shown). Bar, 0.02 substitutions per nucleotide position.

5-5 RAPD method

The Random Amplified Polymorphic DNA (RAPD) method is a PCR based method, comprised of fingerprinting using single or multiple oligonucleotides as the primers [9]. The main aspect of this technique is that, by using a short primer of approximately 10 bases, multiple fragments can be randomly amplified, and the profile of this amplification can determine the pattern of the species or even subspecies. Yamazaki and co-workers (Hokkaido University) utilized this technique for the detection of *A. acidoterrestris*, and have published an account using 3 different primers, to easily differentiate *A. acidoterrestris* from the other *Alicyclobacillus* [10]. This method was affected by the purity and concentration of the DNA and by the type of enzymes and apparatus used and therefore demonstrated poor reproducibility. Further improvements are needed to standardize the method and harmonize these conditions. The method showed a similar performance to the ribotyping analysis, and has proven handy to determine similarity among different strains.

5-6 Amplification of the GyrB gene (*gyrB*)

The *gyrB* is a gene encoding the different topological forms of DNA, and is widely distributed among living organisms. Regarding the nucleotide sequence variability, *gyrB* (average similarity within *Alicyclobacillus* species: 77.2 %) is less conserved than 16S rDNA (average similarity within *Alicyclobacillus* species: 94.5 %), thus it enables a more refined differentiation among the DNA sequences [13]. Just like 16S rDNA, it can be used to determine the nucleotide sequence and identify and differentiate the species. Recently a method using an easier and more specific primer has been developed [11]. Basically the method consists of using the chromosomal DNA extracted from colonies as a template, performing a PCR reaction using a specific primer for the genus *Alicyclobacillus* or *A. acidoterrestris*, and detecting the difference between the reaction products by gel electrophoresis. An example of the method is shown in Fig. 5-7. Positive aspects of this method are the utilization of a low cost, common apparatus in the laboratory and a required time of approximately 5 hours to get the results. The *gyrB* method can also be used for techniques that were developed for 16S rDNA, such as the RT-PCR method [12] and the Lamp (Loop-mediated Isothermal Amplification) method [13]. For details about the RT-PCR method and the Lamp method, please consult the original article [14].

5-7 Ribotyping

This method consists of extracting chromosomal DNA from the colonies, treating them with restriction enzymes, isolating the fragments by gel electrophoresis, transferring them onto a membrane, and performing a southern hybridization with a standard ribosomal DNA. The pattern obtained is image analyzed to establish the relationship among species. This method allows for the identification of the different strains, however, it is difficult to determine the genus and species [13]. In this method, a simple microbial cultivation is prepared and all the subsequent steps can be automated, making the entire process very easy. The required time for analysis is 8 hours, but the equipment (Qualicon Riboprinter-TM, commercialized by Takara Biotech Co. in Japan and by DuPont Qualicon in USA) is relatively expensive and the database is still not well constructed (it is possible to construct your own database). These are some points to be considered and that need to be addressed in future development work.

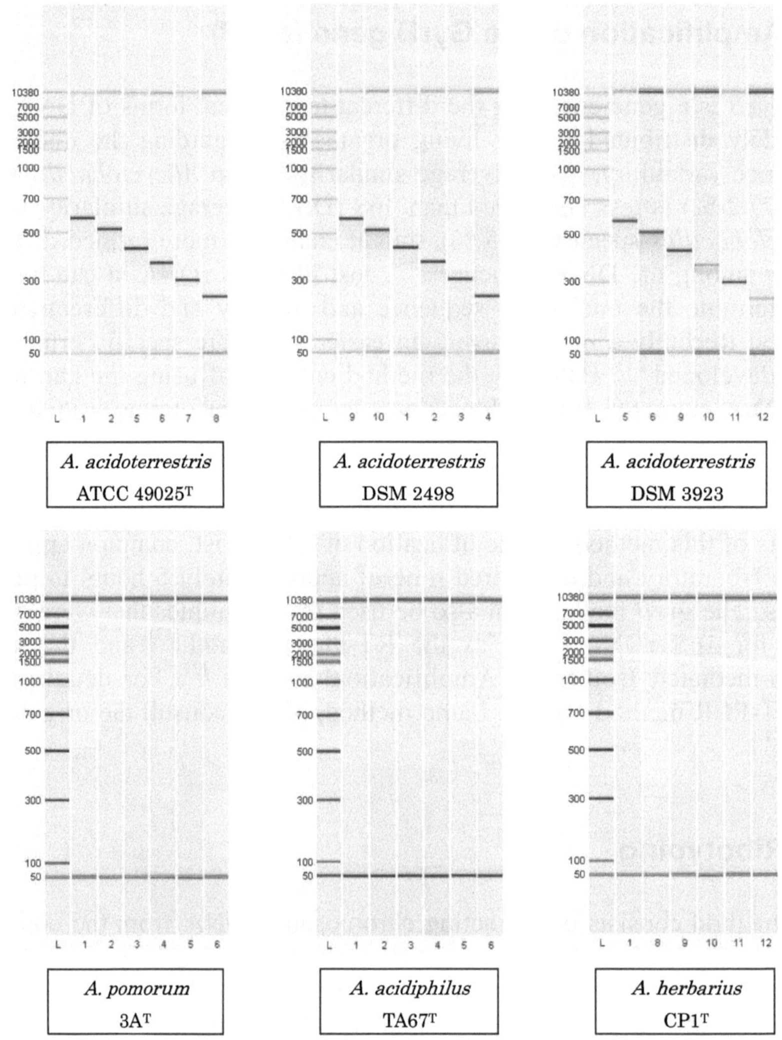

Fig. 5-7. Gel electrophoresis patterns of the PCR products of *Alicyclobacillus* spp. using 6 primer sets specific to *A. acidoterrestris*

M: Marker, Lane 1: Primer set of PF1 & PR2, Lane 2: PF1 & PR1, Lane 3: PF3 & PR2, Lane 4: PF3 & PR1, Lane 5: PF4 & PR2, Lane 6: PF4 & PR1.

5-8 ω-Cyclic fatty acids analysis

This method is based on gas chromatography analysis of the unique ω-cyclic fatty acids, found in *Alicyclobacillus* [3, 15]. Using this method it is possible to identify *Alicyclobacillus* at the genus level. The determination of the chain size of the detected main fatty acid allows grouping at the species level. *A. cycloheptanicus*, *A. herbarius*, *A. kakegawaensis* and *A. shizuokaensis* possess ω-cycloheptil type fatty acid [3, 16] and the others possess the ω-cyclohexil type fatty acid. Exception to this are, *A. pomorum*, *A. contaminans* and *A. macrosporangidus* which do not contain ω-cyclic fatty acids, so they are excluded from detection by this methodology. Other microorganisms that possess ω-cyclic fatty acid are *Curtobacterium* sp. [17], *Propionibacterium* sp. [18] and *Sulfobacillus* spp. [19], but the content of this fatty acid in *Alicyclobacillus* is more than 50 % of the total fatty acids, allowing for an easy differentiation from the other microoorganisms. This method does not permit identification at the species level.

5-9 VIT method

The VIT method (Vermicon Identification Technology, German) is based on the use of gene probes labeled with fluorescent radicals for the detection of the genus *Alicyclobacillus* or specifically *A. acidoterrestris* [20]. Through a selective reaction between these probes and the unknown *Alicyclobacillus*, it is possible to visualize the cells directly under a fluorescence microscope. *A. acidoterrestris* exhibits a red fluorescence and the remaining *Alicyclobacillus* species emit green fluorescence. The positive aspect of this method is that the procedure is simple and when a sufficient number of cells are present, the result is obtained in few hours (in the case of too few cells, an enrichment step is needed).

5-10 Sensory evaluation of guaiacol

This method is based on the metabolic ability of *Alicyclobacillus* species to convert vanillin into guaiacol through vanillic acid. Basically, the target colony is inoculated into a broth (such as YSG broth) supplemented with vanillin, a precursor of guaiacol. The inoculated broth is incubated under appropriated conditions and the presence/absence of guaiacol is determined by sensory analysis [21]. This method is extremely simple but the odor of guaiacol can be masked by the odor of the medium, therefore appropriate training is required.

When YSG broth is used as the basal composition, it is recommended the broth be diluted 2-4 fold with sterile distilled water. This reduces the odor of the yeast extract (it is also recommended to distribute the broth in test tubes for evaluation). The appropriate concentration of vanillin for this evaluation is approximately 1-10 ppm, because the evaluation becomes difficult if the concentration is over or under this range. The precision of the evaluation is increased when the amount of fresh cells is as large as possible, followed by a cultivation for 48-96 h at 50°C.

5-11 Detection of guaiacol

This method determines for the presence of guaiacol producing microorganisms in the sample by qualitative or quantitative determination of guaiacol. The amount of guaiacol in a sample is often very low, thus requiring a method that allows for an accurate determination on the order of 1 ppb. Besides guaiacol, 2,6-dibromophenol or 2,6-dichlorophenol can also be used as indicators [22], however their threshold is around 1000 times lower than guaiacol, and an even more sensitive method is required.

In practice, solid phase microextraction (SPME) and GC-MS can be used, and by selecting an appropriate internal standard, it is possible to detect 0.5 μg/L of guaiacol [23].

5-12 SensiMedia methodology

This method is based on the detection of carbon dioxide (CO_2) produced by *Alicyclobacillus* during their active growth phase. Microbio Co. (Japan) has commercialized a device using liquid media coupled with a carbon dioxide sensor (SensiMedia for *Alicyclobacillus* and SensiMedia for *A. acidoterrestris*). The use of this media allows for differentiation of *A. acidoterrestris*, *A. acidocaldarius* and *A. cycloheptanicus*.

The sample is inoculated into the above media, and if a single living cell is present in the inoculated sample, the carbon dioxide sensor will change from dark blue to transparent and colorless. After cultivation, a specific reagent is added for the differentiation of *A. acidoterrestris*. If positive the color will be light blue or colorless. If negative the color will remain unaltered (blue) [24].

5-13 Assimilation of erythritol

This method is based on information provided by Baumgart in 1997, that *A. acidoterrestris* is the only species among *Alicyclobacillus* species that assimilates erythritol, with acid production [25].

Plates are prepared with BAM medium supplemented with 10 g/L erythritol and 0.015 g/L bromphenol blue as an indicator. Samples (colonies or broth cultures) are streaked onto two plates, one plate being incubated at 46°C and the other at 60°C, for a period of 3 days. If the color surrounding the colonies becomes yellow, the test is positive for *A. acidoterrestris*, and if the color remains unaltered (blue), the result is negative for *A. acidoterrestris*. This test is also possible using liquid media.

Basically, the technique is the same as API 50CH. Since some erythritol-negative strains have been isolated, there is the possibility of a false-negative response.

Keiichi Goto

References

1. Niwa M (2003) *A. acidoterrestris* Rapid Detection Kit. Fruit Processing 13:328-331.
2. Niwa M (2004) Guaiakoru kensyutsu (Peruokishidaze hou) ni yoru sansei inryou yuugai kin no kanri. Japan Food Science 43:23-28 (in Japanese).
3. Goto K, Matsubara H, Mochida K, Matsumura T, Hara Y, Niwa M, Yamasato K (2002) *Alicyclobacillus herbarius* sp. nov., a novel bacterium containing ω-cycloheptane fatty acids, isolated from herbal tea. Int J Syst Evol Microbiol 52:109-113.
4. Goto K, Tanimoto Y, Tamura T, Mochida K, Arai D, Asahara M, Suzuki M, Tanaka H, Inagaki K (2002) Identification of thermoacidophilic bacteria and a new *Alicyclobacillus* genomic species isolated from acidic environments in Japan. Extremophiles 6:333-340.
5. Goto K, Mochida K, Asahara M, Suzuki M, Yokota A (2002) Application of the hypervariable region of the 16S rDNA sequence as an index for the rapid identification of species in the genus *Alicyclobacillus*. J Gen Appl Microbiol 48:243-250.
6. Goto K, Omura T, Hara Y, Sadaie Y (2000) Application of the partial 16S rDNA sequence as an index for rapid identification of species in the genus *Bacillus*. J Gen Appl Microbiol 46:1-8.
7. Goto K, Mochida K, Asahara M, Suzuki M, Kasai H, Yokota A (2003) *Alicyclobacillus pomorum* sp. nov., a novel thermo-acidophilic, endospore-forming bacterium that does not possess ω-alicyclic fatty acids,

and emended description of the genus *Alicyclobacillus*. Int J Syst Evol Microbiol 53:1537-1544.

8. Goto K, Fujita R, Kato Y, Asahara M, Yokota A (2004) Reclassification of *Brevibacillus brevis* strains NCIMB 13288 and DSM 6472 (=NRRL NRS-887)as *Aneurinibacillus danicus* sp. nov. and *Brevibacillus limnophilus* sp. nov. Int J Syst Evol Microbiol 54:419-427.
9. Williams JGK, Kubelik AR, Livak KJ, Rafalski JA, Tingey SV (1990) DNA polymorphisms amplified by arbitary primers are useful as genetic markers. Nucleic Acids Res 18:6531-6535.
10. Yamazaki K, Okubo T, Inoue N, Shinano H (1997) Randomly Amplified Polymorphic DNA (RAPD) for rapid identification of spoilage bacterium *Alicyclobacillus acidoterrestris*. Biosci Biotech Biochem 61:1016-1018.
11. Goto K, Kato Y, Asahara M, Yokota A (2002) Evaluation of the hypervariable region in the 16S rDNA sequence as an index for rapid species identification in the genus *Paenibacillus*. J Gen Appl Microbiol 48:281-285.
12. Yamazaki K, Tezuka H, Shinano H (1996) Isolation and identification of *Alicyclobacillus acidoterrestris* from acid beverages. Biosci Biotech Biochem 60:543-545.
13. Goto K, Mochida K, Kato Y, Asahara M, Ozawa C, Kasai H, Yokota A (2006) Diversity of *Alicyclobacillus* isolated from fruit juices and their raw materials, and emended description of *Alicyclobacillus acidocaldarius*. Microbiol Cult Coll 22:1-14.
14. Shimabukuro C, Inoue T, Nakakita Y, Kaneda H, Shiotani H, Minekawa H, Soejima T (2003) LAMP hou wo mochiita tainetusei kousanseikin no hantei houhou ni tuite. Nihon-Seiryoinryo-Kenkyukai, Dai-13-kai Kenkyu Happyokai (in Japanese).
15. Muramatu T, Niwa M (1992) Inryo seizou ni okeru kanri shihyoukin no doutei to shibousan bunseki ni yoru kanben doutei no kokoromi. J Antibact Antifung Agents 20:637-642 (in Japanese).
16. Deinhard G, Saar J, Krischke W, Poralla K (1987) *Bacillus cycloheptanicus* sp. nov., a new thermoacidophile containing ω-cycloheptane fatty acids. Syst Appl Microbiol 10:68-73.
17. Suzuki K, Saito K, Kawaguchi A, Okuda S, Komagata K (1981) Occurrence of ω-cyclohexyl fatty acids in *Curtobacterium pusillum* strains. J Gen Appl Microbiol 27:261-266.
18. Kusano K, Yamada H, Niwa M, Yamasato K (1997) *Propionibacterium cyclohexanicum* sp. nov., a new acid-tolerant ω-cyclohexyl fatty acid-containing propionibacterium isolated from spoiled orange juice. Int J Syst Bacteriol 47:825-831.
19. Dufresne S, Bousquet J, Boissinot M, Guay R (1996) *Sulfobacillus disulfidooxidans* sp. nov., a new acidophilic, disulfide-oxidizing, gram-positive, spore-forming bacterium. Int J Syst Bacteriol 46:1056-1064.
20. Thelen K, Beimfohr C, Snaidr J (2003) Specific rapid detection of *Alicyclobacillus* by fluorescently-labeled gene probes in fruit juices. Fruit Processing 6:416-418.

21. Kato H, Kawai Y, Tezuka H, Matsunaga M (1993) Sansei inryo no henpai jiko ni kansuru chousa (2) Kounetsusei yugahousaikin niyoru ishu hassei (fureba henka) ni tuite. Kanzumejihou 72:28 (in Japanese).
22. Jensen N, Whitfield FB (2003) Role of *Alicyclobacillus acidoterrestris* in the development of a disinfectant taint in shelf-stable fruit juice. Lett Appl Microbiol 36:9-14.
23. Ono A, Suzuki M, Goto K, Nanjo F (2004) Inshokubutu no ishu genin tonaru guaiakoru no jinsoku na kenshutu hou nituite. Nihon Nougei Kagaku Kai Taikai 2004 nendo taikai kouen youshishu 222 (in Japanese).
24. Kitamura K, Taniyama T, Aoyama F, Mitani M, Matsumoto K, Miyashita M, Ogawa H (2001) Teishoku hou ni yoru ekitai baichi wo mochiita *Alicyclobacillus* zoku saikin no jinsoku kenshutu houhou. Nihon Nougei Kagaku Kai Taikai 1996 nendo taikai kouen youshishu 267 (in Japanese).
25. Baumgart J (1997) *Alicyclobacillus acidoterrestris*: vorkommen, bedeutung und nachweis in gertranken und getrankegrundstoffen. Flüssiges Obst 64:178-180 (in Germany).

Chapter 6: Growth Profile of *Alicyclobacillus* in Fruit Juices

6-1 Introduction

Since the discovery of *Alicyclobacillus* as the odor-producing agent in fruit juice, and with the development of more sensitive and reliable methodologies for its detection, the behavior of *Alicyclobacillus* in fruit juices is gradually becoming clearer. To date, it is known that the growth of *Alicyclobacillus* in fruit juices can be directly or indirectly influenced by temperature, pH and available oxygen (redox potential). In addition, other factors that can impact growth are the type and amount of nutrients present and naturally occurring antimicrobial compounds present in the juice. In this chapter, these factors will be discussed in detail.

6-2 Differences in *Alicyclobacillus* species

Not all species of *Alicyclobacillus* grow in fruit juices under normal environmental conditions, as discussed in Chapter 4. Some species, such as *A. acidoterrestris*, *A. hesperidum* and *A. acidiphilus*, can grow in soft drink products containing fruit juice. Depending on the type of product and the storage temperature (50-60°C) *A. acidocaldarius* and "*A. mali* " may also occasionally grow, although however, the degree of growth is low. The growth of other *Alicyclobacillus* species in soft drinks has not yet been reported.

As presented in Table 6-1 [1], a wide range of differences was recognized in the growth characteristics among strains of *A. acidoterrestris* isolated from various fruit juices. Although all of these strains were identified as *A. acidoterrestris* species at the taxonomic level, some strains grew very well in orange juice but not in lemon juice or grapefruit juice. On the other hand, other strains showed different behavior. This observation of the relationship between the strain and the type of juices in which it can grow is an interesting point from an academic perspective. However, much care must be taken with strains isolated from raw materials or from the industrial environment, especially while their harmfulness is being evaluated in a product.

Table 6-1. Growth of *A. acidoterrestris* in various fruit juices [1)]

Strain	B2065	B2066	B2067	RB1	RB 221	RB 253	RB 346	RB 359	TAB H1
30% apple	+	+	+	-	+	-	+	+	+
30% orange	+	+	-	+	-	+	+	+	+
100% mix-juice	+	+	+	+	+	+	+	+	+
30% lemon	-	+	-	+	+	-	+	+	-
30% grapefruit	-	+	+	+	+	-	+	+	+

Incubation temperature: 35 °C, inoculum level: 10^2 CFU/mL, aerobic condition, incubation period; 2 weeks.
B2065-B2067: isolated from apple juice, RB1, RB253 and RB346: isolated from orange juice, RB221: isolated from isotonic drink, RB359: isolated from mango juice, TAB H1: isolated from pineapple juice. +: Growth, -: no growth.

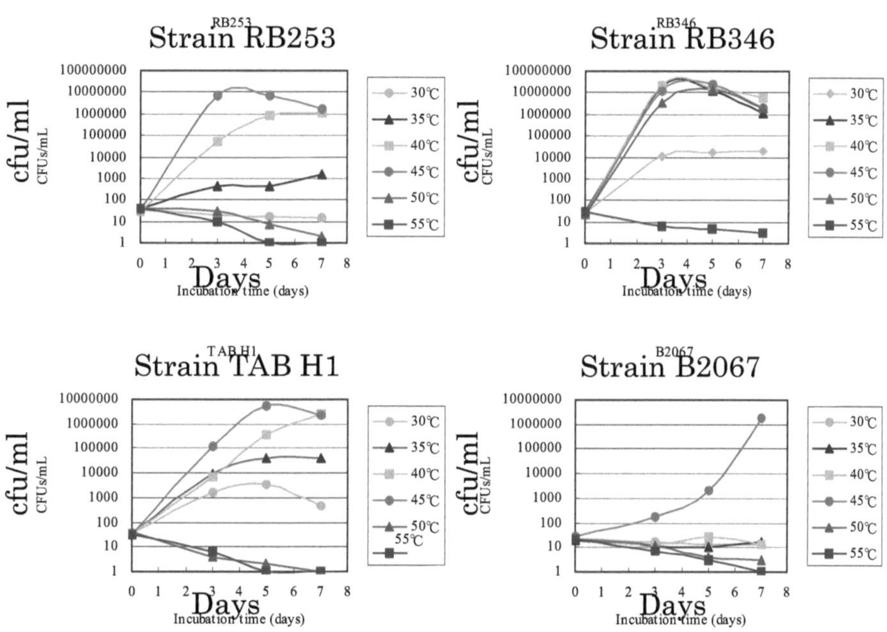

Fig. 6-1. Growth of four *A. acidoterrestris* strains in 100 % orange juice

A. acidoterrestris spores (20-50 CFU/ml) were inoculated into 200 ml of orange juice and then incubated at 30-55 °C for 7 days.

The growth profiles of *A. acidoterrestris* strains at different temperatures are compiled in Fig. 6-1. Growth was not observed for any of the strains at temperatures close to the maximum limit (55°C). However, at lower temperatures great variability among the strains was observed [1]. It is likely that these differences are due to different sensitivities to both temperature and antimicrobial compounds present in the juices. In general, a strain isolated from a fruit juice grows well in that juice.

6-3 Effects of temperature

The temperature for growth of *Alicyclobacillus* species is within a range from 20°C to 70°C. For *A. acidoterrestris*, a species particularly known for its ability to cause undesirable odor (guaiacol) in soft drinks, the temperature range for growth is 20-55°C [1]. In culture media, growth at temperatures closer to the extremes of this range is very slow, requiring up to several weeks to detect observable growth [2]. Similar behavior is also observed when fruit juices is the growth medium. Table 6-2 shows the relationship between the growth of *A. acidoterrestris* and incubation temperature when spores were inoculated into different fruit juices. The results showed no detectable growth was observed in orange, apple or other non-carbonated fruit juices when stored at 4°C. However, when stored at 25, 35 and 45°C, an incremental increase in the cell count of 100 to 10000-fold was observed. With increasing cell numbers odor formation was detected in greater numbers of samples. Guaiacol was detected in all of the spoiled samples [3]. In Japan, the growth profile of a thermo-acidophilic bacterial strain isolated from apple juice was investigated in juices, having from 70-100 % apple content. As shown in Fig 6-2, growth was observed at 35 and 45°C within a short period of time. However, growth was not detected at 25°C even after storage for 20 days [4]. This suggests that if the products are kept at room temperature, the possibility of deterioration is very low, in the presence of this *Alicyclobacillus* strain. Some authors have reported that in orange, grape and grapefruit juices growth was not detected even at 35 and 45°C. It is well known that the high polyphenol content in grape and apple juice, does inhibit the growth of *Alicyclobacillus*. This will be discussed in further detail later. Reports of growth in orange and other fruit juices have become increasingly common, yet as shown in Table 6-3, the variation in growth behavior in fruit juices has been shown to be very dependent on the *Alicyclobacillus* strain used. Therefore it must be kept in mind that the results of investigational studies may be greatly influenced by the strain utilized in these studies.

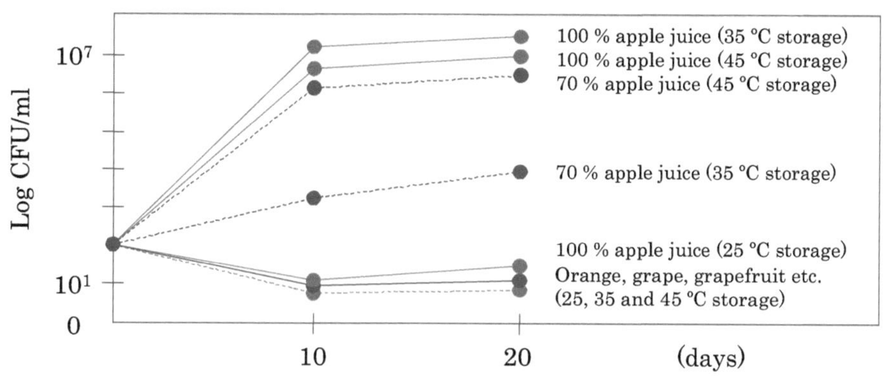

Fig. 6-2. Growth pattern of *A. acidoterrestris* in various fruit juices [5)]

The inoculated strain derived from apple juice.

6-4 Effects of the redox potential and available oxygen

Alicyclobacillus are strictly aerobic bacteria and so we can predict their growth in fruit juice will be clearly affected by the amount of dissolved oxygen and the redox potential of the juice.

As shown in Fig. 6-3 the growth characteristics of *A. acidoterrestris* are greatly modified when ascorbic acid is added to apple juice resulting in a lower redox potential; the growth of the bacteria is completely inhibited at a concentration of more than 15 mg/100 mL ascorbic acid [5)].

Dissolved oxygen in the juice is greatly affected by the oxygen concentration in the headspace of the filled container. As shown in Fig. 6-4, when fruit juices (apple and orange) are bottled without leaving any headspace or, alternatively, when the headspace is replaced by a N_2/CO_2 gas mixture with 0.84 % residual oxygen, the growth of *A. acidoterrestris* is completely inhibited in both cases. The effect of the headspace oxygen level on the growth profile of *A. acidoterrestris* differs with juice type. In the case of orange juice, growth was still observed with 0.1 % residual oxygen, as shown in Fig. 6-5. However, the fact that growth remains controlled with 0.84 % oxygen concentration, as shown in Fig. 6-4, is believed to be due to CO_2 acting as a bacteriostat in a synergistic mode of action. On the other hand, in the case of apple juice and white grape juice, 0.1 % residual oxygen suppresses the growth of the bacteria [5]. There is also a recent report on the effect of intermittent shaking headspace and temperature on the growth of *A. acidoterrestris* in apple juice [14].

Table 6-2. The growth of *Alicyclobacillus acidoterrestris*, detection of taint and production of guaiacol in orange juice, apple juice and non-carbonated fruit juice-containing drinks [3]

Produced	Temperature (°C)	Incubation time (day)	Count CFU/ml	Sensory tainted	Guaiacol (ppb)
Orange juice	4	0	100	No	NT
		21	600	No	NT
	25	0	100	No	NT
		1	8×10^3	No	NT
		6	6×10^5	Yes	14-1
	35	0	100	No	NT
		1	2×10^4	No	NT
		6	1×10^6	Yes	13-3
	44	0	100	No	NT
		1	3×10^4	No	NT
		6	6×10^6	Yes	1-2
Apple juice	4	0	100	No	NT
		21	<100	No	NT
	25	0	100	No	NT
		1	8×10^3	No	NT
		6	6×10^4	No	NT
		10	2×10^7	Yes	11-6
	35	0	100	No	NT
		1	6×10^4	No	NT
		6	3×10^6	Yes	17-3
	44	0	100	No	NT
		1	3×10^4	No	NT
		6	2×10^7	Yes	33-7
Non-carbonated fruit juice-containing drink	4	0	100	No	NT
		13	60	No	NT
	35	0	100	No	NT
		3	1×10^5	Yes	32-3
	44	0	100	No	NT
		3	1×10^6	Yes	100-8

NT: not tested.

The type of container material also affects the dissolved oxygen in the juice. Therefore it is easily deduced that the type of material used for the container also affects growth. The growth characteristics of *A. acidocaldarius* (the strain was described as *A. acidocaldarius* in the reference [6], but it might actually be *A. acidoterrestris* because this strain produced guaiacol) inoculated in Valencia orange juice in a paper carton, incubated at 35°C, were observed. There was very weak growth when the paper pack was internally coated with aluminum, but deterioration was observed within 2 days in the absence of this coating. On the other hand, slight growth was detected when the juice was packaged in aluminum cans, but the growth never exceeded 100-fold, and there was no alteration in appearance, odor or pH value. Since oxygen permeation was almost non-existent in the situations described, it is thought that only the oxygen remaining after the filling process can be used for growth of the microorganisms.

Table 6-3. Growth pattern of *A. acidoterrestris* in beverages [9]

	A. acidoterrestris strain								
	B2065	B2066	B2067	RB1	RB221	RB253	RB346	RB359	TAB H1
30% apple juice	+	+	+	-	+	-	+	+	+
30% apple juice	+	+	-	+	-	+	+	+	+
30% lemon juice	-	+	-	+	+	-	+	+	-
30% grape fruit juice	-	+	+	+	+	-	+	+	+
100% mix fruit juice	+	+	+	+	+	+	+	+	+

B2065-B2067: isolated from apple juice, RB1, RB253 and RB346: isolated from orange juice, RB221: isolated from isotonic drink, RB359: isolated from mango juice, TAB H1: isolated from pineapple juice. +: Growth, -: no growth.

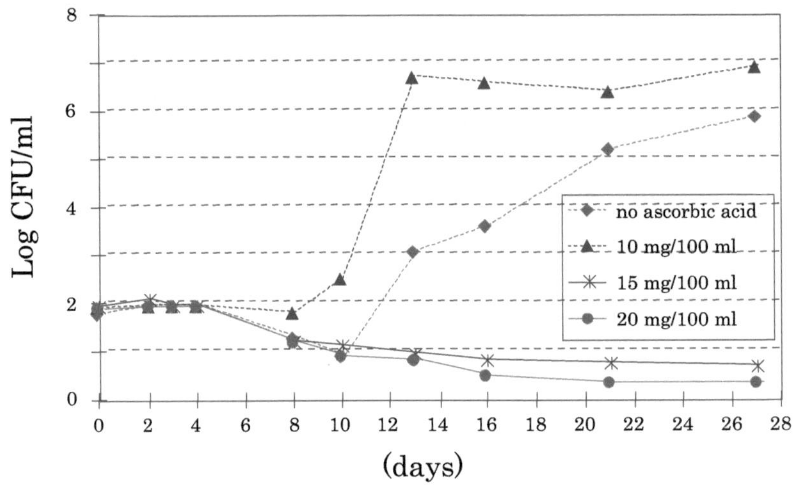

Fig. 6-3. Growth pattern of *A. acidoterrestris* DSM 2498 in ascorbic acid added apple juices under aerobic condition (35 °C) [5)]

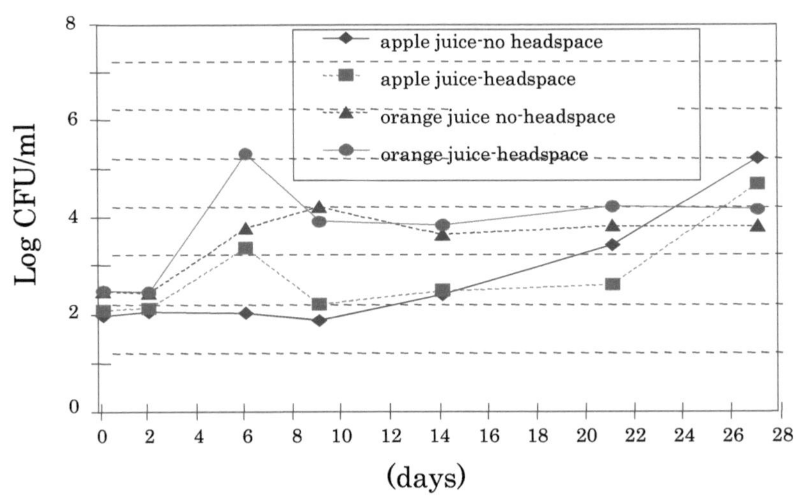

Fig. 6-4. Growth pattern of *A. acidoterrestris* DSM 2498 in various filling conditions (30 °C) [5)]

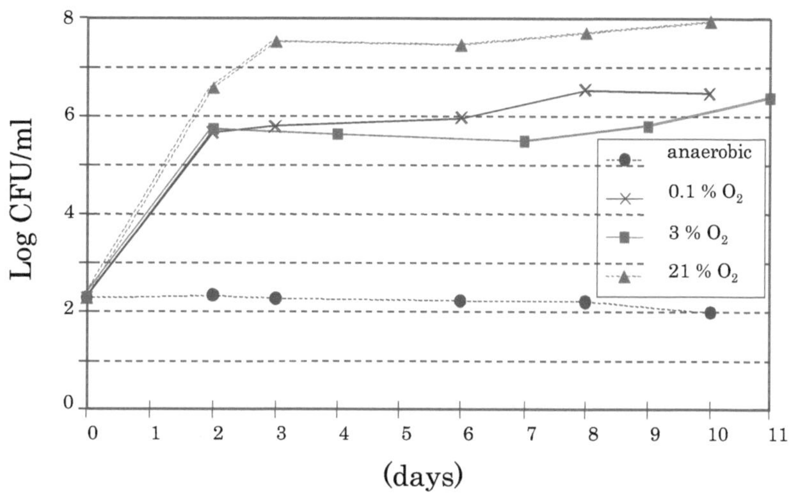

Fig. 6-5. Growth pattern of *A. acidoterrestris* DSM 2498 in various O_2 conditions (35 °C) [5]

6-5 Effects of the composition of the fruit juice and the origin of the raw material

The growth behavior of *Alicyclobacillus* species is greatly affected by the juice type. The growth characteristics of *Alicyclobacillus* species were compared in different types of fruit juices, as shown in Table 6-4 [7]. The growth was most easily observed in tomato, followed by apple and orange juices. The pH of tomato juice is approximately 4, and in the case of the other juices the pH is approximately 3.5, a value near the lower limit of the range for *Alicyclobacillus*. The lower solids content in tomato juice may also impact growth in this juice. Growth was not detected in the following mixed juices: apple/orange/pineapple, apple/raspberry/grape and cranberry cocktail. This may be because these juices have a very low pH, of 2.5-3.0. The same type of study was carried out with *A. acidoterrestris* [8], and these results are presented in Table 6-5. The growth of *A. acidoterrestris* is strongly affected by the source of juice, with growth being observed in grapefruit, apple, orange, pear, white grape and tomato; but not in apple-cranberry, 10 % fruit juices , pineapple juice and salsa. This could be attributed to the possibility that inhibitory compounds exists in some of the components of these juices and inhibit either germination of the spores or growth of the cells.

Table 6-4. Growth of heat-activated VF and WAC spores in various commercial fruit-juice beverages [7]

Juice	Brix%	pH	Growth* Log N/N_0	
			VF	WAC
Apple, brand A	11.4	3.5	1.3	1.8
brand B	11	3.5	1.1	1.6
Apple-grape-cherry blend	14.8	3	-1.4	-0.9
Apple-orange-pineapple	14.8	2.9	-0.1	0.3
Apple-rasberry-grape	12.2	2.8	-1.5	-1.1
Apple-red grape	12.4	3.7	-0.4	-1
Granberry cocktail	14	2.4	-1.4	-1.3
Grape red, brand A	15.8	3.3	-0.7	-0.7
brand B	13.6	2.9	-1.2	-1.5
Grape-apple-cherry blend	12.4	3.7	0.7	-0.8
Grapefruit	10.4	3.2	0.3	-1.3
Orange	12	3.6	1.2	-0.8
Pineapple	13.4	3.3	0.4	-0.7
Pineapple-orange	12.8	3.4	0.3	-0.6
Prune	18.8	3.7	-0.6	-0.1
Tomato	7	4	2.3	2.4
Tropical fruit blend	13.6	3.7	1.2	-1.3

*: Two-days incubation at 43 °C.

Table 6-5. Spoilage of juices by *A. acidoterrestris* [8]

Product	Initial pH	Initial Brix%/total solubule solids	Growth
Apple cranberry juice	2.67	14.2	-
10% fruit juice	2.97	11.8	-
Grapefruit juice	3.28	10.0	+ (slowly)
Apple juice	3.47	11.6	+
Pineapple juice	3.61	13.0	-
Orange juice	3.66	11.4	+
Pear juice	3.82	12.4	+
White grape juice	3.87	16.2	+
Salsa	3.92	12.5 TSS	-
Tomato juice	4.27	6.1	+

Incubation at 35 °C, inoculum levels of 10^1 to 10^4 spores/ml.

The Brix of the juices is another factor that affects the growth of *Alicyclobacillus*. When a range of Brix is obtained by adding water or sugar into Riesling grape juice, at a 5-20 % level, a 100-fold growth difference was observed when incubated at 43°C for 2 days. At a Brix of 21.6 %, the growth was completely inhibited [7, 9].

Investigation of the *Alicyclobacillus* behavior in different juices, has shown growth of *Alicyclobacillus* in white grape juice and inhibition in red grape juice, having the same Brix and pH (Table 6-6) [7, 9]. This inhibition is attributed to the presence of neutral phenolic compounds. In Table 6-7, the inhibitory effect of several neutral phenolic compounds on *Alicyclobacillus* is listed. A strong inhibition is observed with catechin gallate where the growth is inhibited with 1000 mg/L [7]. A variety of polyphenols extracted from grapes were also investigated for their inhibitory effect on *A. acidoterrestris*, and the results indicated cumaric acid and ferulic acid had a strong inhibitory affect with a MIC (minimal inhibitory concentration) of 150-200 mg/L [10].

Table 6-6. Growth of VF and WAC spores in diluted juice of various grape cultivars [7]

	Brix%	pH	Growth* Log N/N_0	
			VF	WAC
White juices				
Riesling	10.8	3.4	2.0	1.9
Seyval	9.5	3.2		2.2
Chardonnay	11.3	3.3	-0.9	-1.0
Elvira	7.8	3.4	2.2	2.2
Cayuga White	8.6	2.8	1.8	0.17
Red juices				
Early Burgundy	9.1	3.5		-0.8
Gamay Noir	10.2	3.1		-0.9
Gamay Beaujolais	10.3	3.3	-1.2	-0.72
Pinotage	10.9	3.8	-1.0	-0.57
Cabernet Sauvignon	12.2	3.7	-1.0	-0.63
Concord	10.0	3.5	-1.2	

*: Two days at 43 °C.

Table 6-7. Growth of spores inocula in 6°Brix apple juice containing different concentrations of added neutral phenolic compounds [7)]

Phenolic	mg/l	Growth* Log N/N_0	
		VF	WAC
Catechin	100	1.7	2.3
	200	1.7	2.3
	1000	1.9	2.9
Epicatechin	100	1.7	2.3
	200	1.8	2.3
	1000	1.9	2.6
Catechin-gallate	100	1.8	2.1
	200	1.8	2.1
	1000	-1.1	-0.8
Quercetin	100	1.9	2.2
	200	1.8	2.2
	1000	1.8	2.6

*: Two days at 43 °C.

With regard to ethanol, growth is inhibited when the concentration is above 6 %. This indicates *Alicyclobacillus* cannot grow in normal table wines, and there is no risk of their deterioration [11)].

The effect of organic acids on the growth characteristics of *A. acidocaldarius* are presented in Fig. 6-6, 6-7 and 6-8. Citric acid and malic acid did not show a significant inhibition of growth, however the other acids were inhibitory especially at lower pH. The decreasing order of inhibition was acetic acid, adipinic acid, lactic acid, fumaric acid. Acetic acid was found to strongly inhibit growth [12)].

With regard to NaCl, a 2 % solution was not sufficient to control growth, but control of growth of *Alicyclobacillus* was reported to ocur when increased to 5 % NaCl [13)].

Hajime Tokuda

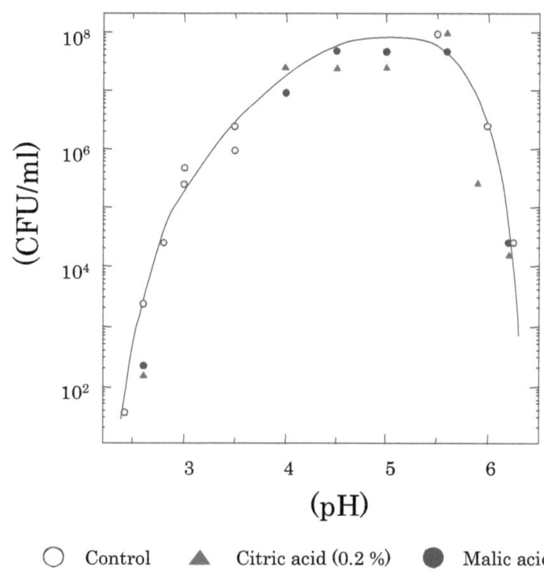

Fig. 6-6. Effects of pH and addition of citric and malic acids on the growth and/or germination of spores of *A. acidocaldarius* AC-1 [12)]

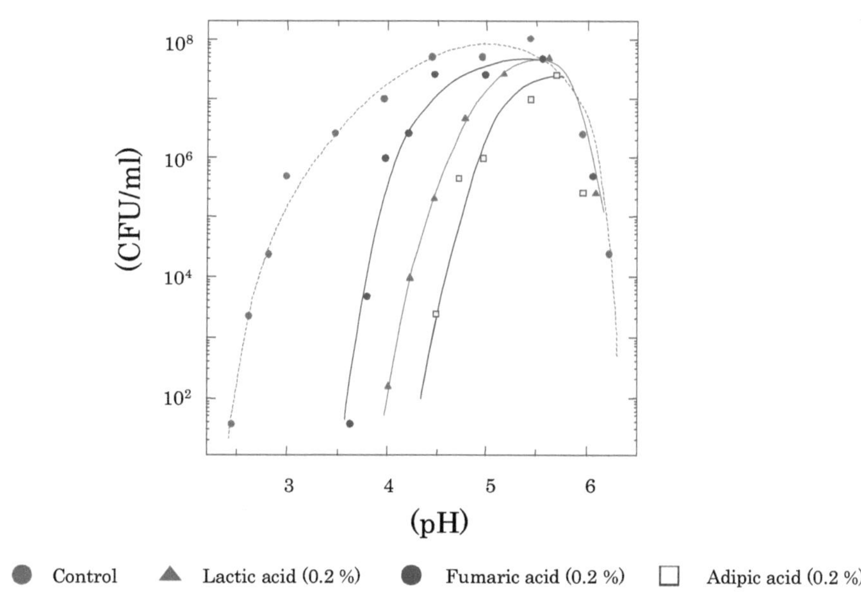

Fig. 6-7. Effects of pH and addition of lactic, and fumaric and adipic acids on the growth and/or germination of spores of *A. acidocaldarius* AC-1 [12)]

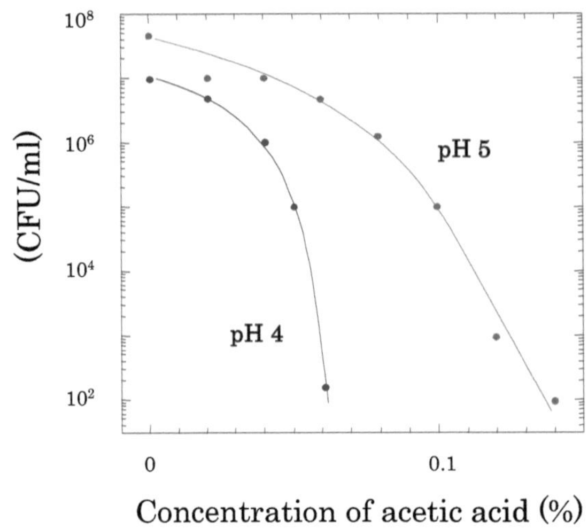

Fig. 6-8. Effects of pH and addition of acetic acids on the growth and/or germination of spores of *A. acidocaldarius* AC-1 [12)]

References

1. Goto K (2003) Annals of the ILSI Japan Thermoacidophilic Bacteria Symposium. "Saishin tainetsusei kousansei kin no kisochishiki to sono taisakuho" (in Japanese).
2. Goto K (2000) Thermoacidophilic spore-forming bacteria: *Alicyclobacillus*. J Antibacterial Antifungal Agents 28:13-38 (in Japanese).
3. Pettipher GL, Osmundson ME, Murphy JM (1997) Methods for the detection and enumeration of *Alicyclobacillus acidoterrestris* and investigation of growth and production of taint in fruit juice and juice-containing drinks. Lett Appl Microbiol 24:185-189.
4. Niwa M, Hirota Y, Shirasu Y (1991) Yunyu kaju yori kenshutu sareta shinki tainetsusei kousansei kin. Nihon-Kaju Kyokai Hou 9:31-42 (in Japanese).
5. Cerny G, Duong HA, Hennlich W, Miller S (2000) *Alicyclobacillus acidoterrestris*: influence of oxygen content on growth in fruit juices. Food Australia 52:289-291.
6. Ikegami Y, Enda M, Matsui C, Nakagawa K. (1996) Barenshia orenji jusu chu ni okeru kousansei kin no zoushoku. Toyo Shokuhin Kogyo College/Toyo Shokuhin Laboratories, Kenkyu Hokokusho 21:117-122 (in Japanese).

7. Splittstoesser DF, Churey JJ, Lee CY (1994) Growth characteristics of aciduric sporeforming bacilli isolated from fruit juice. J Food Protection 57:1080-1083.
8. Walls I, Chuyate R (2000) Spoilage of fruit juices by *Alicyclobacillus acidoterrestris*. Food Australia 52:286-288.
9. Splittstoesser DF, Lee CY, Churey JJ (1998) Control of *Alicyclobacillus* in the juice industry. Dairy Food Environ Sanitation 18:585-587.
10. Oita S, Koyama N (2002) Antibacterial effect of grape polyphenols against thermoacidophilic bacteria *Alicyclobacillus acidoterrestris*. Nippon Shokuhin Kagaku Kogaku Kaishi 48:555-558 (in Japanese).
11. Splittstoesser DF, Churey JJ (1996) Unique spoilage organisms of musts and wines. In Wine Spoilage Microbiology Conference, (Toland, T., and Fugelsang, K. C. ed.), pp.36-41, California State University, Fresno.
12. Ikegami Y, Enda M, Matsui C, Nakagawa K (1996) Kaju inryo kara bunri sareta kousansei kin ni tsuite. Toyo Shokuhin Kogyo College/Toyo Shokuhin Laboratories, Kenkyu Hokokusho 21:105-115 (in Japanese).
13. Yamazaki K, Teduka H, Shinano H (1996) Isolation and identification of *Alicyclobacillus acidoterrestris* from acidic beverages. Biosci Biotech Biochem 60:543-545.
14. Michelle W, Carol AP (2005) The effect of intermittent shaking headspace and temperature on the growth of *Alicyclobacillus acidoterrestris* in stored apple juice. Int J Food Sci Tech 40:557-562.

Chapter 7: Distribution and Epidemiology

Improvement in the detection of *Alicyclobacillus* species has lead to many more reports of the isolation of *Alicyclobacillus* from various environments, including spoiled fruit juices. The accumulation of these case studies has provided us with more extensive epidemiological information. In the course of investigating the cause of such cases and identifying appropriate measures to eliminate them, we have come to a better understanding of the characteristics of *Alicyclobacillus* with regard to beverage production. This chapter describes the epidemiology of *Alicyclobacillus*, in relation to their distribution in soil environments (primary distribution), then the secondary distribution in fruit juice or the manufacturing environment.

7-1 Primary distribution

The origin of *Alicyclobacillus* isolates is considered primarily to be from soil [1)]. As expected from their thermophilic and acidophilic features, they are often isolated in acidic soils in volcanic areas and hot springs [2-8)]. However, reports of their isolation in the Antarctic [9, 10)] indicates that their distribution is worldwide. Table 7-1 shows the locations in Japan where *Alicyclobacillus* has been isolated from soil and water samples. Many *Alicyclobacillus* isolates have been reported in acidic environments around hot springs and volcanic zones throughout Japan, from the north to the south. Phylogenetic analysis using 16S rDNA sequences identified all as *Alicyclobacillus acidocaldarius* and related species. However, some morphological and genetic differences identified regional varieties. *Alicyclobacillus* strains have also been isolated from soils with neutral pH [11, 12)]. For example, 50 strains were obtained by an attempt to isolate "spore-forming thermophilic and acidophilic bacilli" from various neutral-pH soil samples. Among them, 23 strains were isolated under a combination of high temperature and acidic pH conditions. They all possessed ω-cyclic fatty acids, the characteristic cell components in the *Alicyclobacillus* [12)]. Their characteristic features are summarized in Table 7-2.

Table 7-1. Strains and their sources [2)]

Source		Samples	Strains
Soil	Iozan, Hokkaido pref.	1	
Soil, Water	Lake Goshiki, Hokkaido pref.	1	7 (FCS strain)
Soil, Water	Norikura, Nagano pref.	7	
Soil, Water	Arashiyama, Kyoto pref.	3	
Soil	Yubara, Okayama pref.	1	
Soil, Water	Yanahara, Okayama pref.	10	
Soil	Okayama Univ., Okayama pref.	20	
Soil	Yuba, Okayama pref.	1	
Soil	Sauna bath, Okayama pref.	1	
Activated sludge	Treatment plant, Okayama pref.	3	6 (EMG strain)
Soil, Water	Shionoe, Kagawa pref.	5	
Soil, Water	Unzen, Nagasaki pref.	25	1 (UZ strain)
Soil, Water	Minami Aso, Kumamoto pref.	10	10 (SAS strain)
Soil, Water	Mt. Aso, Kumamoto pref.	3	
Soil, Water	Amagase, Oita pref.	7	
Soil, Water	Sujiyu, Oita pref.	29	27 (KH strain)
Soil, Water	Ebino, Miyazaki pref.	6	
Soil, Water	Kirishima, Kagoshima pref.	15	9 (MIH strain)

Table 7-2. Properities of *Bacillus acidocaldarius* and thermo-acidophilic bacilli isolated from soil [12)]

Strains	B.a.	ND2A	MD3A	MD2A	GK2B	B/50	B/45
pH-range of growth on:							
Complex medium	4-7	3-4	-	2-4	3-4	3-5	3-7
Semi-synthetic medium	4-5	3-5	3-5	3-4	3-5	3-5	3-6
Temperature range	26-62	26-70	22-62	22-62	22-70	36-80	26-70
Temperature for good growth	36-53	36-53	36-53	36-53	36-53	45-70	36-53
Growth on carbon sources:							
Glucose	+	+	+	+	+	+	+
Saccharose	+	+	+	+	+	+	+
Glycerol	+	+	+	+	+	-	+
Gluconate	+	+	+	+	+	+	+
Sorbitol	+	+	-	+	+	+	+
Inositol	+	-	-	-	-	+	+
Starch	+	-	+	-	-	+	+
Acetylmethylcarbinol	+	-	-	-	-	-	+
Location of the oval spore in the slightly swollen sporangium	t-s	c-s	t	t	c-s	c-s	t-s

t: Terminal, s: subterminal, c: central, B.a: *Bacillus acidocaldarius* (*Alicyclobacillus acidocaldarius*).
Temperatures used were: 22, 26, 36, 45, 53, 62, 70 and 80 °C.

7-2 Secondary distribution

Awareness that *Alicyclobacillus* is responsible for incidences of an off-odor in fruit juices has prompted many investigations aimed at elucidating the route of contamination. The fruit, that is the raw material of the fruit juice (leaves and the surface of the fruit), the orchard (soil etc.), the fruit processing plant environment, the intermediate product from each stage of the production process, the equipment and water used in the process, have all been monitored [13, 14]. In 1996 and 1997, the technical committee of The Brazilian Association for Citrus Exporters (ABECitrus) carried out an extensive survey on *Alicyclobacillus* distribution [15]. Three major orange processing factories in São Paulo State and suburban orange farms around each factory were selected for the studies. Testing for *Alicyclobacillus* was conducted on one hundred samples taken from normally harvested fruits, fruit on ground, harvested fruit in baskets, orange leaves, ground soils, sprinkler water at orchards, fruit before and after the storage in silos, before and after the chlorine wash, the washing water after the chlorine wash, processed orange juice before concentration, juice after pasteurization, concentrates (final products adjusted with recovered aroma) as well as recovered orange aroma and flavoring. The results are shown in Tables 7-3, 7-4 and 7-5. Relatively large numbers of *Alicyclobacillus* were detected in all soil samples at a level of 10^4-10^6 cfu/g. A small amount was detected on the surface of the leaves and fruit, which is not surprising since soil may easily come in contact with and adhere to the surface of the fruit and leaves in the orchard. The contamination tended to be higher in dry seasons than in rainy seasons, due it is thought to the dry soil being blown up by the wind in the dry season, and washed away by the rain in the rainy season. This data will serve as an important guide for developing strategies for the control of *Alicyclobacillus* during orange processing; namely how we can reduce the contamination on fresh fruits can be reduced, and to what extent *Alicyclobacillus* contamination during the juice processing can be eliminated. Since *Alicyclobacillus* are highly heat-resistant they are not pasteurized by the regular heat treatment performed on squeezed juices at orange factories. In particular, the efficacy of the wash and the degree of contamination by *Alicyclobacillus* from the washing water is regarded to greatly influence the extent to which the final product is microbe free. This study indicates we have reason for concern since some samples of the washing water showed relatively high levels of *Alicyclobacillus* contamination. Strict elimination of processing of fruits that have dropped on the ground is strongly recommended in order to prevent heavy contamination by *Alicyclobacillus* from the orchard soils.

Table 7-3. Counting and detection of *Alicyclobacillus* sp. in the first sampling (June-July, 1996) [13)]

Sample description	Code	North	Code	Central	Code	South
		Counts (enrichment)		Counts (enrichment)		Counts (enrichment)
Fruit from tree (var. Hamlin)	1a	12 CFU/kg (+)	42a	2.02×10^2 CFU/kg (+)	18a	6 CFU/kg (+)
Fruit on the ground (var. Hamlin)	2a	2 CFU/kg (+)	43a	$>6 \times 10^2$ CFU/kg (+)	20a	2 CFU/kg (+)
Fruit pooled in bags (var. Hamlin)	3a	12 CFU/kg (+)	44a	2 CFU/kg (-)	22a	8 CFU/kg (+)
Leaves from tree (var. Hamlin)	4a	1.33×10^2 CFU/kg (+)	45a	33.3 CFU/kg (+)	26a	1.0×10^2 CFU/kg (+)
Soil from orange groves (var. Hamlin)	5a	4 × 104 CFU/kg (+)	46a	2.9×10^6 CFU/kg (+)	24a	1×10^5 CFU/kg (+)
Irrigation/spray water (var. Hamlin)	-	ND	47a	6.4 CFU/mL (ND)	-	ND
Fruit from tree (var. Pera)	70a	$>6 \times 10^2$ CFU/kg (+)	48a	<10 CFU/kg (+)	66a	$>6 \times 10^2$ CFU/kg (-)
Fruit on the ground (var. Pera)	71a	3.7×10^2 CFU/kg (+)	49a	4 CFU/kg (+)	67a	28 CFU/kg (-)
Leaves from tree (var. Pera)	72a	$4,3 \times 10^2$ CFU/kg (+)	50a	2×10^2 CFU/kg (+)	68a	33.3 CFU/kg (+)
Soil from orange groves (var. Pera)	73a	4×10^4 CFU/kg (+)	51a	1.12×10^7 CFU/kg (+)	69a	7.5×10^6 CFU/kg (+)
Fruit after the silo	6a	6 CFU/kg (+)	52a	6 CFU/kg (-)	28a	10 CFU/kg (+)
Washed fruit (before chlorination)	7a	1.54×10^2 CFU/kg (+)	53a	10 CFU/kg (+)	29a	4 CFU/kg (-)
Washed fruit (after chlorination)	8a	2.84×10^2 CFU/kg (+)	54a	4 CFU/kg (+)	30a	2 CFU/kg (-)
Water from wash system (from noozles)	9a	1.5×10^6 NMP/mL (ND)	55a	2.2×10^2 CFU/mL (ND)	31a	<1 CFU/mL (ND)
Condensed water (from the evaporator)	-	ND	56a	<1 CFU/mL (+)	41a	<1 CFU/mL (ND)
Water from essential citrus peel oil recovery system	-	ND	62a	<10 CFU/mL (+)	-	ND
Fresh juice (before evaporator)	10a	<10 CFU/mL (+)	57a	1.7×10^3 CFU/mL (+)	32a	6.2×10^2 CFU/mL (+)
Concentrated juice (at evaporator outlet)	11a	20 CFU/ mL (+)	58a	5.3×10^3 CFU/mL (+)	33a	<10 CFU/mL (+)
Concentrated juice in the blending tank (before finishing)	12a	10 CFU/ mL (+)	-	ND	34a	2×10^2 CFU/mL (-)
Concentrated juice (after finishing)	-	ND	-	ND	35a	4.3×10^2 CFU/mL (+)
Citrus pulp cells (after finisher)	13a	3×10^2 CFU/ g (+)	60a	2×10^2 CFU/g (+)	39a	50 CFU/g (+)
Pulp (discharge from centrifuge)	14a	<10 CFU/ g (+)	61a	8×10^2 CFU/g (+)	40a	20 CFU/g (+)
Essential citrus peel oil	15a	<10 CFU/ mL (ND)	74a	20 CFU/mL (ND)	36a	20 CFU/mL (ND)
Orange aroma (water phase)	16a	<10 CFU/ mL (ND)	-	ND	-	ND
Orange essence oil (oil phase)	17a	<10 CFU/ mL (ND)	-	ND	-	ND

ND: Not determined.

Table 7-4. Counting and detection of *Alicyclobacillus* sp. in the second sampling (September-October, 1996) [13)]

Sample description	Code	North Counts (enrichment)	Code	Central Counts (enrichment)	Code	South Counts (enrichment)
Fruit from tree (var. Pera)	80	6 CFU/kg (+)	100	92 CFU/kg (+)	120	8 CFU/kg (+)
Fruit on the ground (var. Pera)	81	76 CFU/kg (+)	101	10 CFU/kg (+)	121	10 CFU/kg (-)
Fruit from the truck (var. Pera)	82	2 CFU/kg (+)	102	16 CFU/kg (+)	122	2 CFU/kg (+)
Leaves from tree (var. Pera)	83	33.3 CFU/kg (+)	103	$>10^3$CFU/kg (+)	123	33.3 CFU/kg (+)
Soil from orange groves (var. Pera)	84	4×10^5 CFU/kg (+)	104	2×10^4 CFU/kg (+)	124	7×10^5 CFU/kg (+)
Fruit after the silo	85	$>6 \times 10^2$ CFU/kg (+)	105	2 CFU/kg (+)	125	10 CFU/kg (+)
Washed fruit (after chlorination)	86	18 CFU/kg (+)	106	8 CFU/kg (+)	126	24 CFU/kg (+)
Water from wash system (noozles)	87	17 CFU/mL (ND)	107	43 NMP/mL (ND)	127	15 CFU/mL (ND)
Condensed water (from the evaporator)	93	0.7 CFU/mL (ND)	115	1.3 CFU/mL(ND)		
Water from essential citrus peel oil recovery	95	55 CFU/mL (ND)	117	1.8×10^2 CFU/mL (+)	137	60 CFU/mL (+)
system (day 1/day 7)	96	8.6×10^2 CFU/mL (+)	119	97 CFU/mL (+)	138	<10 CFU/mL (+)
Fresh juice (before evaporator)	88	5×10^2 CFU/mL (+)	108	50 CFU/mL (+)	128	10 CFU/mL (-)
Concentrated juice in the blending tank (before finishing)	89	3×10^2 CFU/mL (+)	109	1×10^2 CFU/mL (+)	129	80 CFU/mL (+)
Concentrated juice (after finishing)	90	3×10^2 CFU/mL (+)	110	70 CFU/mL (+)	130	1.2×10^2 CFU/mL (+)
Citrus pulp cells (after finisher)	91	6×10^2 CFU/g (+)	111	2.5×10^3 CFU/g (+)	131	1.5×10^2 CFU/g (+)
Essential citrus peel oil	92	1×10^2 CFU/mL (ND)	113	<10 CFU/mL (ND)	133	<10 CFU/mL (ND)
Orange aroma (water phase)	98	2.8×10^2 CFU/mL (ND)			-	ND
Orange essence oil (oil phase)	99	70 CFU/mL (ND)			-	ND
Water (industrial/potable)	97	0.24 CFU/mL (ND)	118	0.3 CFU/mL(ND)	139	0.2 CFU/mL (ND)
Water (evaporate from waste)	94	0.35 CFU/mL (ND)	116	93 NMP/mL(ND)		

ND: Not determined.

Table 7-5. Counting and detection of *Alicyclobacillus* sp. in the third sampling (December, 1996-January, 1997) [13)]

Sample description	Code	North	Code	Central	Code	South
		Counts (enrichment)		Counts (enrichment)		Counts (enrichment)
Fruit from tree (var. Valencia)	197	6 CFU/kg (+)	175	<1 CFU/kg (+)	219	2 CFU/kg (+)
Fruit on the ground (var. Valencia)	198	<1 CFU/kg (+)	176	6 CFU/kg (+)	220	4 CFU/kg (+)
Fruit pooled in trucks (var. Valencia)	199	2 CFU/kg (+)	177	2 CFU/kg (+)	221	<1 CFU/kg (-)
Leaves from tree (var. Valencia)	200	10^2 CFU/kg (+)	178	33.3 CFU/kg (+)	222	66.6 CFU/kg (+)
Soil from orange groves (var. Valencia)	201	1 × 105 CFU/kg (+)	179	3×10^4 CFU/kg (+)	223	1.6×10^5 CFU/kg (+)
Fruit from tree (var. Pera)	152	2 CFU/kg (-)	156	>6×10^2 CFU/kg (+)	160	<1 CFU/kg (-)
Fruit on the ground (var. Pera)	153	2 CFU/kg (-)	157	>6×10^2 CFU/kg (+)	161	2 CFU/kg (+)
Leaves from tree (var. Pera)	151	33.3 CFU/kg (-)	155	33.3 CFU/kg (-)	159	33.3 CFU/kg (+)
Soil from prange groves (var. Pera)	150	2.1×10^3 CFU/kg (+)	154	6×10^4 CFU/kg (+)	158	1.6×10^5 CFU/kg (+)
Fruit after the silo	202	64 CFU/kg (+)	180	10 CFU/kg (-)	224	18 CFU/kg (+)
Washed fruit (after chlorination)	203	96 CFU/kg (+)	181	12 CFU/kg (+)	225	<1 CFU/kg (+)
Water from wash system (noozles)	204	6×10^3 CFU/mL (ND)	182	1×10^3 CFU/mL (ND)	226	0.1 CFU/mL (ND)
Condensed water (from the evaporator)	211	<1 CFU/10 mL (ND)	189	3 CFU/mL (ND)	233	<1 CFU/10 mL (ND)
Water from essential citrus peel oil recovery	213	1.4×10^2 CFU/mL (ND)	191	60 CFU / mL (+)	235	10 CFU/mL (+)
system (at day 1 and at day 7)	214	2.8×10^2 CFU/mL (+)	192	6×10^2 CFU / mL (+)	236	<1 CFU/mL (-)
Fresh juice (before evaporator)	205	2.18×10^2 CFU/mL (+)	183	1.87×10^2 CFU/mL (+)	227	16 CFU/mL (+)
Concentrated juice in the blending tank (before finishing)	206	5.4×10^3 CFU/mL (+)	184	5×10^2 CFU/mL (+)	228	10 CFU/mL (+)
Concentrated juice (after finishing)	207	2.26×10^2 CFU/mL (+)	185	3.4×10^3 CFU/mL (+)	229	1.8×10^2 CFU/mL (+)
Citrus pulp cells (after finisher)	208	10 CFU/g (+)	186	10 CFU/g (+)	230	90 CFU/g (+)
Essential citrus peel oil	210	<10 CFU/mL (ND)	188	10 CFU/mL (+)	232	10 CFU/mL (ND)
Orange aroma (water phase)	216	<1 CFU/mL (ND)	-	ND	239	1.1×10^2 CFU/mL (ND)
Orange essence oil (oil phase)	217	<1 CFU/mL (ND)	-	ND	-	ND
Water (industrial/potable)	215	<1 CFU/mL (ND)	-	ND	237	<1 CFU/10 mL (ND)
Water (evaporate from waste)	212	10 CFU/mL (ND)	190	<1 CFU/10 mL (ND)	234	1 CFU/mL (ND)

ND: Not determined.

7-3 Taxonomy and epidemiological information on *Alicyclobacillus* in food ingredients

Epidemiology of currently known *Alicyclobacillus* species, such as *A. acidocaldarius*, *A. acidoterrestris*, *A. hesperidum* and *Alicyclobacillus* genomic species 1 is summarized in Table 7-6 [16, 17]. Among them, *A. acidocaldarius*, *A. acidoterrestris* and *Alicyclobacillus* genomic species 1 are the most frequent isolates from soil. *A. acidocaldarius* strains are con-

Table 7-6. Source of *Alicyclobacillus* species [16)]

Name	Source	Frequency of detection	Population in Japan
A. mail	Apple juices	Rare	Yes
Alicyclobacillus genomic species 1	Soil, fruits, syrup, etc.	High	Yes
A. sendaiensis	Soil	Rare	Yes
Alicyclobacillus genomic species 2	Soil	Rare	Yes
A. acidocaldarius subsp. *acidocaldarius*	Soil, fruits, syrup, etc.	High	Yes
A. acidocaldarius subsp. *rittmannii*	Soil	Rare	No
A. acidoterrestris	Soil, fruits, syrup, etc.	Low	No
A. acidiphilus	Raw materials of beverages	Rare	No
A. hesperidum	Soil	Rare	No
A. pomorum	Beverages	Rare	No
Alicyclobacillus sp. (New Species)	Fruits	Rare	No
A. cycloheptanicus	Soil	Rare	No
A. herbarius	Herb leaves	Rare	No

*Only information officially reported was described.
**Cases in which the original source was from overseas, even if thereafter identified in Japan, are shown as "No".

firmed not only from acidic environments in hot springs and volcanoes but also from plant samples, including fresh fruits and processed juices, and thus is the most widely recognized *Alicyclobacillus* species. *A. acidoterrestris* has also been isolated quite widely from soils and fruits, as well as fruit juice products and their processing water at factories [11, 12, 18-23]. As such, it is the most common *Alicyclobacillus* responsible for spoilage of fruit juices and acidic foods.

As stated, precautions against the contamination of food and beverages by *Alicyclobacillus* species should be taken because they are found widely in soil and agricultural products. Special notice should be given to the fact that *A. acidocaldarius*, *A. acidoterrestris* and *Alicyclobacillus* genomic species 1 have been detected not only in fruits but commonly in liquid sugars as well. Fructose, glucose, sucrose, high fructose corn syrups, and pectins from both citrus and apples, are reported as sources of *A. acidoterrestris* and other species. Judging from the overall distribution and characteristics of the *Alicyclobacillus* species, it can be concluded that they are able to survive in a variety of agricultural products for which strong heat pasteurization is not used for taste quality reasons. Moreover, attention should be paid to the fact that contamination by *Alicyclobacillus* spores can occur not only in juices but also in other raw materials and additives used for food products and beverages. Unfortunately though, there is insufficient epidemiological information at this time on *Alicyclobacillus* in food ingredients other than fruit juices. Further extensive studies and surveys in cooperation with producers and organizations concerned are strongly advised if we are to fully understand and deal with current issues regarding the extensive contamination of raw materials by *Alicyclobacillus*.

Hajime Tokuda

References

1. Goto K (2000) Kouonsei kousansei gahou keisei saikin: *Alicyclobacillus* zoku saikin. J Antibact Antifung Agents 28:499-508 (in Japanese).
2. Goto K, Tanimoto Y, Tamura T, Mochida K, Arai D, Asahara M, Suzuki M, Tanaka H, Inagaki K. (2002) Identification of thermoacidophilic bacteria and a new *Alicyclobacillus* genomic species isolated from acidic environments in Japan. Extremophiles 6:333-340.
3. Belly RT, Brock TD (1974) Widespread occurrence of acidophilic strains of *Bacillus coagulans* in hot springs. J Appl Bact 37:175-177.

4. Takayanagi S, Sugimori K, Sendo M (1986) Sanseisen ni seisoku suru biseibutu no doutei to bunpu. Toho Univ General Rev 18:33-43 (in Japanese).
5. Uchino F, Doi S (1967) Acido-thermophilic bacteria from thermal waters. Agr Biol Chem 31:817-822.
6. Darland G, Brock TD (1971) *Bacillus acidocaldarius* sp. nov., an acidophilic thermophilic spore-forming bacterium. J Gen Microbiol 67:9-15.
7. Tanimoto Y, Inagaki K, Tanaka H (1996) Kounetusei kousanseikin no tansaku, tanri oyobi shoseishitu no kentou. Okayama Univ. Academic Rev 85:15-21 (in Japanese).
8. Hiraishi A, Inagaki K, Tanimoto Y, Iwasaki M, Kishimoto N, Tanaka H (1997) Phylogenetic characterization of a new thermoacidophilic bacterium isolated from hot springs in Japan. J Gen Appl Microbiol, 43, 295-304.
9. Hudoson JA, Daniel RM, Morgan HW (1989) Acidophilic and thermophilic *Bacillus* strains from geothermal heated Antarctic soil. FEMS Microbiol Lett 60:279-282.
10. Nicolaus B, Improta R, Manca C, Lama L, Esposite E, Gabacorta A (1998) Alicyclobacilli from an unexplored geothermal soil in Antarctica: Mount Rittmann. Polar Biol 19:133-141.
11. Deinhard G, Blanz P, Poralla K, Altan E (1987) *Bacillus acidoterrestris* sp. nov., a new thermotolerant acidophile isolated from different soils. System Appl Microbiol 10:47-53.
12. Hippchen B, Roll A, Poralla K (1981) Occurrence in soil of thermo-acidophilic bacilli possessing ω-cyclohexane fatty acids and hopanoids. Arch Microbiol: 129:53-55.
13. Eguchi SY, Mnfio GP, Pinhatti MEC, Azuma EH, Variane S, Colecao de Culturas Tropical (CCT) Fundacao Andre Tosello and ABECitrus (1999) Detection methods, ecology and involvement in the deterioration of fruit juices study.
14. Borlinghaus A, Engel R (1997) *Alicyclobacillus* incidence in commercial apple juice concentrate (AJC) suppliers-method development and validation. Fruit Processing 7:262-266.
15. Wisse CA, Parish ME (1998) Isolation and enumeration of sporeforming, thermoacidophilic, rod-shaped bacteria from citrus processing environments. Dairy Food Environ Sanitation 18:504-509.
16. Goto K (2003) Tainetsusei kousansei kin (*Alicyclobacillus* zoku saikin) no kensa kanbetsuhou ni tsuite. Nihon Kaju Kyokai (in Japanese).
17. Goto K (2003) Saishin tainetsusei kousansei kin no kiso chishiki to sono taisakuhou. Tainetsusei kousansei kin shinpojiumu ILSI JAPAN (in Japanese).
18. Yamazaki K, Isoda C, Tezuka H, Kawai H, Shinano H (1997) Sansei inryo henpai kin *Alicyclobacillus acidocaldarius* no tainetsusei tosono seigyo. Nihon Shokuhin Kagaku Kougaku Kaishi 44:905-911 (in Japanese).
19. Suzuki T (1989) Sansei inryo kanzume no kousan kin ni tuite. Shokuhin to Youki 30:503-506 (in Japanese).

20. Komitopoulou E, Boziaris IS, Davices EA, Delves-Broughton J, Adams MR (1999) *Alicyclobacillus acidoterrestris* in fruit juices and its control by nisin. Int J Food Sci Tech 34:81-85.
21. Eiroa MNU, Junqueira VCA., Schmidt, FL (1999) *Alicyclobacillus* in orange juice: Occurrence and heat resistance of spores. J Food Protection 62:883-886.
22. Splittstoesser JD, Churey J, Lee C (1994) Growth characteristics of spore-forming bacilli isolated from fruit juices. J Food Protection 57:1080-1083.
23. McIntyre S, Ikawa JY, Parkinson N, Haglund J, Lee J (1995) Characteristics of an acidophilic *Bacillus* strain isolated from shelfstable juices. J Food Protection 58:319-321.

Chapter 8: Factors of spoilage caused by *Alicyclobacillus* and prevention measures

To date, many *Alicyclobacillus* isolates have been derived mainly from soil and hot springs [1-4]. However, since a major spoilage incident of apple juice by *Alicyclobacillus* in Germany, 1982 [5], the bacteria has been isolated from various food grade fruit juices [6,-12]. *Alicyclobacillus* contamination of fruit juices is typically introduced via soil residue remaining on the fruit surface [13]. Similarly, *Alicyclobacillus* contamination of refined sugar is introduced from raw plant material [14, 15]. *Alicyclobacillus* contaminations are not limited to these mentioned above [16, 17], and they pose serious problems for the food industry because adaptive countermeasures are not presently in place. In this chapter, reasons for spoilage due to *Alicyclobacillus* and effective preventative measures as reported so far are discussed with regard to materials, production, products and distribution.

Kanjiro Takahashi

8-1 Factors and prevention of spoilage (contamination) in materials

The Brazilian Association for Citrus Exporters (ABECitrus) conducted comprehensive studies into the cause of orange juice contamination due to *Alicyclobacillus*, as a cooperative research project of member companies [18]. The research targets included examining soils in the groves, the production processes, orange juice products, and also the improvement of guidelines (The project report is available from http://www.abecitrus.com.br/.). The relationship between soil and fruit with regard to contamination by *Alicyclobacillus* is reported as below:

"In all grove soil sample analyses so far performed, we encountered relatively high counts of *Alicyclobacillus*, in the range of 10^4-10^6 cfu/g, and occurrences were regardless of regional differences between North Central and South regions of the São Paulo State orange belt (Tables 7-3 to 7-5). *Alicyclobacillus* was also found on the surface of fruits and leaves in the grove, with counts in the range of <10-10^2 cfu/kg product. On the contrary, fruit and leaf samples which were collected immediately after a rainy day or in the rainy season (third sampling; Table 7-5) had significantly lower counts of *Alicyclobacillus*, which suggested that the main contamination source of the factory is the fruit itself (ultimately, the soil) and enough washing at the initial phase of the process prior to concentration of juice can help reduce *Alicyclobacillus* counts. Furthermore, *Alicyclobacillus* was detected in the irrigation water (Table 7-3) and washing water (Tables 7-3 to 7-5), which points to water as one of the most important control points [19]."

Since the above report was published, Brazilian juice producers have taken special care not to use ground fruits contaminated with soil particles, and to thoroughly wash fruit with well processed water [or by using bactericide-treated water, see 8-2-2(3)] as prevention measures against *Alicyclobacillus*. These countermeasures are effective and practical for easily washable fruits like oranges and apples. However, no effective methods have been reported thus far for pineapples, herbs, and the raw materials of sugar such as corn and beets.

8-2 Reasons for spoilage in production, and spoilage prevention measures.

8-2-1 Reasons for spoilage in production, and spoilage prevention measures.

As mentioned previously, the source of *Alicyclobacillus* contamination is soil. Therefore, contamination occurs throughout the production lines when raw materials (fruits etc.) that are carrying soil are subsequently processed.

In order to reduce *Alicyclobacillus* contamination of plant material in the factory, the raw material must firstly be washed thoroughly. Since bacterial counts cannot be reduced sufficiently by washing with just cold or warm water, detergents or bactericides are added to the washing water [see 8-2-2(3)]. It is preferable to use water that has not been contaminated with *Alicyclobacillus*, but if contaminated water must be used then it should first be processed by elimination or sanitization. Special attention must be taken to ensure treatment chemicals do not remain in the products. The fruit-washing step is the most important control point, as it precedes all downstream processing steps. For raw materials that are difficult (or impossible) to wash, it is consequently very difficult (or impossible) to control *Alicyclobacillus*. All production processes including extraction, filtration, centrifugation, concentration, filling, pipelines will be contaminated by *Alicyclobacillus* that is introduced on plant material. To prevent *Alicyclobacillus* contamination in each step of the process, it is essential to carry out adequate and regular washing and sanitization programs. Moreover, appropriate selection of methods and chemicals to be used for washing and sanitization must be made. Use of alkaline (for removal of dirt, proteins, oils etc.) and acid (for removal of inorganic and sanitization) solutions together with detergents is effective in the elimination of *Alicyclobacillus* during washing and sanitization. Careful selection of chemicals suitable to the particular production line is necessary [see 8-2-2(3)]. In addition to such washing and sanitization programs, devices that do not allow the temperature of the in-process product to exceed 20°C are useful to prevent *Alicyclobacillus* from remaining and growing. However, as *Alicyclobacillus* spores are extremely stable even at low temperatures, it is also recommended to periodically wash and sanitize equipment that is maintained under freezing point.

As spores of *Alicyclobacillus* are heat-resistant, it is difficult to kill them in the sanitization process by the standard procedures [see 8-2-2(2)] without product deterioration. Therefore, for products that can be filtrated, filter elimination technologies are more effective. For products that cannot be filtrated, sanitization technologies using pressure, ultraviolet rays, or radiation are reportedly effective [see 8-2-2(1)]. Thus, as a means to prevent against *Alicyclobacillus* contamination in production, maximization of the efficiency and effectiveness of washing and sanitization processes is vitally important, and due consideration should be given to determining the washing program best suited to the practicalities of each production site. The topic of washing sanitization is discussed in detail in "Science and Technology of Washing Sanitization (Science Forum)" and "Biofilm (Science Forum)".

Keiichi Goto

8-2-2 Details of prevention measures

Unlike *E. coli*, there are no standard regulations for *Alicyclobacillus* control. Consequently, industries and corporations have been determining their own course of action [20]. In Brazil, as explained before, ABECitrus is playing a key role in promoting comprehensive investigations on *Alicyclobacillus*. According to an ABECitrus report, it is considered that spoilage due to *Alicyclobacillus* can be reduced to a negligible level by making the inspection methods widely known, decontaminating plants, cooling quickly after extraction, and storing products at 30°C or below [21]. Therefore, we will further describe here aspects of sanitization, decontamination, and use of bactericides and antibacterial agents as prevention measures against *Alicyclobacillus*.

(1) Sanitization

Methods which have been reported to kill bacterial spores include heating, ultra-high pressure, radiation, ultraviolet and ozone. Heating, pressure, radiation and ultraviolet will be discussed here, as they have been reported as effective for sanitization of *Alicyclobacillus*. The conditions stated below should be regarded as an example only since they depend on species, strains, growth and environmental parameters.

(1)-1 Heat sanitization

Effects of ultra-high temperature (UHT) against *Alicyclobacillus* are shown in Table 8-1. The heat sanitizer is further categorized into two types: plate and tube, based on the heat interchanger. Spores of *A. acidoterrestris* can be killed in clear apple juice by treatment at 112°C × 60 sec or 120°C × 20 sec heating with a plate type, or by 112°C × 60 sec or 120°C × 15 sec heating with a tube type heat sanitizer.

Table 8-1. Effect of UHT pasterization on survival of *Alicyclobacillus* spores [22]

Juices	Type	Pasterization condition	*A. acidoterrestris*	*A. acidocaldarius*
Apple (clear) (50 Brix%)	Plate	112 °C × 60 sec	-	+
		120 °C × 15 sec	+	+
		120 °C × 20 sec	-	+
	Tube	112 °C × 60 sec	+	+
		120 °C × 15 sec	-	+
Orange (60 Brix%)	Plate	112 °C × 60 sec	-	+
		112 °C × 95 sec	-	-
		120 °C × 10 sec	-	+
		120 °C × 15 sec	-	-
	Tube	112 °C × 60 sec	-	-
		120 °C × 15 sec	-	+

'-: not surviving, +: surviving.

(1)-2 Sanitization by pressure

Methods that utilize pressure can be used as non-heating sanitization methods. Examples that utilize pressure together with heating or emulsifier are introduced here.

a) Simultaneous use of pressure and heat

It is reported that while *A. acidoterrestris* spores in apple juice can survive a pressure treatment (621 MPa × 10 min), they are killed by simultaneous treatment with high pressure and heating (45-90°C × 1-10 min) (Table 8-2).

Table 8-2. Inhibitory effects of high pressure and heat on *A. acidoterrestris* spores in apple juice [23)]

Pressure (MPa)	Heat condition	Log reduction (D)*
0	90 °C × 1 min	0
207	45 °C × 10 min	>3.5
	71 °C × 1 min	>3.5
	90 °C × 5 min	>5.5
414	71 °C × 1 min	>4.0
	71 °C × 10 min	>5.5
	90 °C × 1 min	>5.5
621	22 °C × 10 min	0
	71 °C × 1 min	>4.0
	71 °C × 10 min	>5.5
	90 °C × 1 min	>5.5

*: Log (Initial) - Log (After treatment).

b) Simultaneous use of pressurization and emulsifier

While bactericidal effects improve when heating is combined with high pressure treatment, it is reported that a sporicidal effect of 5.5 D against *A. acidoterrestris* spores can be obtained by the addition of 0.045 % emulsifier (monolaurin, sucrose fatty acid ester), even under relatively mild heating conditions of 45°C (Table 8-3).

(1)-3 Radial pasteurization

Gamma-ray, X-ray and electron beam can be used for irradiation of foods and food packaging materials. However, in Japan, the potato is the only food that is permitted to be irradiated (for the inhibition of tuber germination).

Table 8-3. *Alicyclobacillus* spore inhibition and inactivation in juice by pressure, chemical preservatives, and mild heat [24)]

Strain	Medium	L1695 (%)*1	Log reduction (D)*2
N1098	Apple juice	0	2
		0.025	2
		0.040	2.5
		0.045	5.5

Pressure treatment: 392 MPa × 10 min × 45 °C.
*1: Chemical preservatives (Sucrose laurate).
*2: Log (Initial) - Log (After treatment).

Radio sensitivity of *A. acidoterrestris* GD3B strain is almost constantly about 1.0-1.2 kGy of the D_{10} value, independent of the culture conditions (in saline at pH 6.8, in citrate buffer at pH 4.0, or on filter paper) and type of radiation (gamma-ray or electron beam). In the citrate buffer at pH 4.0, 95°C, the two decimal reduction time of *A. acidoterrestris* was about 6 min (from an initial count was 10^6-10^7). After this period a marked decrease in heat sensitivity was observed and heating for up to 60 min did not improve efficacy. However, it is reported that when the bacterium is first treated with 1.0 to 1.5 kGy radiation, subsequent heating at 95°C × 30 min almost completely kills 10^6 spores [25)].

(1)-4 Ultraviolet pasteurization

Ultraviolet pasteurization is generally used for aseptic treatment of food packaging and filling pasteurization of food surfaces. It is also used for the sterilization of microbes adhering to packaging materials, and the air and water used in food factories. Cases of *Alicyclobacillus* contamination are few, but it has been reported that the D value of the juice drink isolate *A. acidocaldarius* AC-1 spores for UV irradiation was 150 μW·min/cm^2, which was almost equivalent to that of *B. subtilis* spores whose D value was 185 μW·min/cm^2 [26)].

(2) Filter elimination

Alicyclobacillus spp., especially *A. acidoterrestris* is not sensitive to the standard bactericidal methods used for acidic drinks [27)]. Therefore, filter elimination technology is promising for the removal of *Alicyclobacillus* spores from materials that are able to be filtrated.

(2)-1 Membrane filter

Filter elimination equipment fitted with membrane filters with 0.45 μm or smaller pore size are effective for eliminating *Alicyclobacillus* spores in production processes of soft drinks and related materials (liquid sugars, juices etc.) (Table 8-4).

Table 8-4. Removal of *A. acidoterrestris* spores from syrup or soft drinks [28)]

Filter brand (Pore size rating)	Spores suspended medium (10^4-10^5 CFU/ml)		
	68 % sucrose syrup	75 % fructose syrup	20 % apple juice
SCW800® (0.8 μm)	+	+	+
EKW® (0.65 μm)	+	-	-
EBW® (0.45 μm)	-	-	-

+: Positive, -: Negative.

(2)-2 Zeta Plus filter

Alicyclobacillus spores in apple juice can be eliminated using the 60H type of "Zeta Plus Filter" which is a depth-type filter with zeta potential and density gradient (Table 8-5). The zeta plus filters are not easily clogged, which increases the volume which can be processed. It is particularly suitable for filter-processing highly viscous liquids, highly contaminated fluids and gel-like solutions.

Table 8-5. Removal of *A. acidoterrestris* spores by Zetaplus filter [29)]

Zetaplus filters	Log reduction value*1 (D)	Spores (CFU)	
		Initial (900 ml)*2	After treatment
30H (0.6-1 μm)	3.5	5.4×10^6	1.8×10^3
60H (0.2-0.5 μm)	>6.5	3.3×10^6	0

*1: Log (Initial) - Log (After treatment), *2: Apple juice.

(3) Bactericides

The main synthetic bactericides used in food products and in food manufacturing environments are shown in Table 8-6 [30-39)]. Ethanol is treated not as a food additive but as a food by the Food Sanitation Law. Sodium hypochloride, slightly acidic electrolyzed water, and hydroxyperoxide are designated as bactericides, and can be used for sanitization of foods. But according to the Food Sanitation Law, slightly acidic electrolysed water and hydroxyperoxide are prohibited from being contained in foods. With regard to the bactericidal efficiency for thermophilic acidophilic bacteria, general bactericides are explained at the top of the table, followed by bactericidal preparations.

Table 8-6. Efficacy of disinfectants in killing spores [30-39]

Category	Disinfectant	Concentration	Temp. (°C)	Time (min)	Spores	Log reduction* (D)
Alcohol	Ethanol	30-80 %	20	5	*B. cereus*	0
	Sodium hypochlorite	200-500 ppm (pH 6)	20	5	*B. subtilis*	4
Chloride	Slightly acidified hypochlorous acid water	15 ppm (pH 5-6.5)	30	10	*B. subtilis*	2.5
	Chlorine dioxide	1 ppm (pH 7)	20	10	*B. subtilis*	3
	Hydrogen peroxide	10%	25	30	*B. subtilis*	4
Peroxide	Peracetic acid	400 ppm	40	10	*B. cereus*	7
	Ozone	5 ppm (liquid)	15	1.3	*B. subtilis*	4
		20 ppm (gas)	Room Temp.	2	*B. subtilis*	100%
Cation surfactant	Benzalukoniumu chloride	0.05-0.5%	-	-	-	0

*: Log (Initial) - Log (After treatment).

(3)-1 Chlorine compounds

Chlorine compounds are categorized into hypochlorites, chlorine dioxides and organic chlorines. Sodium hypochlorite of hypochlorites, chlorine dioxide and sodium chlorite of chlorine dioxides are introduced as follows (Table 8-6).

(a) Sodium hypochlorite (NaOCl)

Sodium hypochlorite, one of the hypochloritic bactericides, is used as a bactericide in food manufacturing factories because of its strong oxidative effects. Sanitization effects of hypochlorites are pH dependant. However, as the pH range of slightly acidic water is much narrower than that of hypochlorites (pH 5.0 to 6.5), sanitization is hardly affected by pH. And furthermore, as chlorine loses its sanitization effect in the presence of organic compounds, it is important to thoroughly wash the objects to be sanitized, such as food manufacturing equipment and fruits prior to sanitization. Sanitization effects of chlorine are stronger with higher temperatures. Because sodium hypochlorite has 4 D sanitization effect against *B. subtilis* spores at 200-500 ppm of effective chlorine concentration at pH 6.0 (Table 8-6), it is also expected to have a sanitization effect on thermophilic acidophilic bacteria. A sanitization effect of 6 D or more is achieved when 1000 ppm sodium hypochlorite (NaOCl) solution is contacted with *A. acidoterrestris* spores for 10 min at 23°C (Table 8-7). It is reported that K-agar (pH 3.7) is more suitable for the detection of *A. acidoterrestris* spores than OSA (Orange Serum Agar, pH 5.0) or potato dextrose agar (pH 3.5). A 1.3 D sanitization effect is obtained when 1000 ppm sodium hypochlorite (NaOCl) solution is contacted with *A. acidoterrestris* spores on the surface of an apple at 23°C for 1 min (Table 8-8). It is considered that the insufficient sanitization effect is due to the natural cutin (a wax-like substance surrounding the plant surface) that exists on the surface of apple fruits making it hydrophobic.

(b) Chlorine dioxide (ClO_2)

Chlorine dioxide bactericides can be produced by the reaction of sodium chlorite ($NaClO_2$) and Cl_2. Chlorine dioxide is a strong oxidant and characteristically remains-in water longer than chlorine because it reacts poorly with ammonium and nitrogen compounds [34]. The stabilized chlorine dioxide preparation which is produced by dissolving ClO_2 gas into purified water, with addition of soda stabilizers and adjustment of pH to 8.5, can be stored for about one year [41].

Chlorine dioxide can provide a sanitization effect against *A. acidoterrestris* of 4.4 D, and 5.4 D or more at 40 ppm and 50 ppm, respectively, when used to treat spores for 5 min. A 4.8 D or greater sanitization effect against *A. acidoterrestris* spores the surface of apples is achieved by 40 ppm × 4 min or 120 ppm × 1 min contact with chlorine dioxide (Table 8-9) [42]. Chlorine dioxide is used in juice producing factories in Brazil because it does not produce the harmful byproducts that sodium hypochlorite can release.

Chlorine dioxide (about 100 ppm) can be used to destroy microbes in water that is produced during the concentration process of orange juice. This recycled water is only reused after confirming that chlorine dioxide does not remain [43].

Table 8-7. Efficacy of disinfectants in killing spores of *A. acidoterrestis* [40]

Disinfectants	pH	Concentration	Log reduction (D)*
		200 ppm	2.4
NaOCl	6.9 ± 0.2	500 ppm	5.6
		1000 ppm	>6.0
		500 ppm	0.4
$NaClO_2$	2.5	850 ppm	0.9
		1200 ppm	1.8
		1 %	0.4
H_2O_2	4.4 ± 0.2	2 %	5.5
		4 %	>5.6

Treatment: 10 min × 23 ± 2 °C, medium: K agar.
*: Log (Initial) - Log (After treatment).

Table 8-8. Efficacy of disinfectants in killing spores of *A. acidoterrestis* on apples [40]

Disinfectants	pH	Concentration	Log reduction (D)
NaOCl	6.9 ± 0.2	500 ppm	0.5
		1000 ppm	1.3
$NaClO_2$	2.5	1200 ppm	0.4
H_2O_2	4.4 ± 0.2	2%	0.1

Treatment: 10 min × 23 ± 2 °C, medium: K agar.

(c) Sodium chlorite ($NaClO_2$)

Sodium chlorite is a chlorine dioxide bactericide (Table 8-6). The sanitization effect obtained by contacting *A. acidoterrestris* spores with 1200 ppm sodium chlorite ($NaClO_2$) solution at 23°C for 10 min is only 1.8 D (Table 8-7). Similarly, there is almost no sanitization effect (0.4 D) against *A.*

acidoterrestris spores on the surface of apple fruits by contacting 1200 ppm sodium chlorite solution for 1 min at 23°C. (Table 8-8).

Table 8-9. Efficacy of chlorine dioxide in killing spores of *A. acidoterrestis* [42]

	Concentration (ppm)	Time (min)	Log reduction (D) *
Medium	40	0.5	0.1
		1	0.7
		5	4.4
	80	0.5	1.0
		1	2.8
		5	>5.4
	120	0.5	2.1
		1	4.8
		5	>5.4
Apples	40	1	1.5
		2	3.2
		3	4.5
		4	>4.8
	120	1	>4.8

*: Log (Initial) - Log (After treatment).

(3)-2 Hydroperoxide

Hydroperoxide is a peroxide sanitizer (Table 8-6). Peroxide has a sanitization effect due to its strong oxidizing power. Although it has been widely used as a sanitizing and bleaching agent of foods, following its recognition as a carcinogen, the food sanitation law has been revised to ban residues of hydroperoxide remaining in food. While hydroperoxide has a sanitization effect on a wide range of microorganisms, it must be used at considerably high concentrations and high temperatures to be effective in a short time. Unlike sodium hypochlorite there is almost no effect of pH on the sanitization effects of hydroperoxide [44]. It is widely used to sanitize the surface of packaging materials in the food industry.

A 5.5 D sanitization effect can be obtained by contacting 2 % hydroperoxide (H_2O_2) solution with *A. acidoterrestris* spores for 10 min at 23°C (Table 8-7). However, there was no sanitization effect observed when 2 % hydroperoxide (H_2O_2) was contacted with *A. acidoterrestris* spores on the surface of apple fruits at 23°C (Table 8-8).

(3)-3 Sanitizing preparations

(a) Oxonia Active

Oxonia Active (Ecolabo) is a sanitizing preparation composed of peracetic acid, hydroperoxide, acetic acid, and stabilizers [37]. Peracetic acid is similar to hydroperoxide, a peroxide sanitizer, however it has far stronger sanitizing power than hydroperoxide.

Depending on the strain, its sanitization effect on *Alicyclobacillus* spores can be 4.3 D or more when it is contacted at 1 % concentration at 50°C, for 15 min (Table 8-10).

Table 8-10. Efficacy of peracetic acid disinfectants in killing spores of *Alicyclobacillus* [45)]

Disinfectants	Conc. (%)	Temp. (°C)	Time (min)	Log reduction (D)*1		
				strain A*2	strain B*2	*B. cereus* spores
Oxonia Active	1	50	0.5	0.7	>4.3	
			2	0.9	>4.3	
			5	3.4	>4.3	
			15	>4.4	>4.3	
		40	10			<1
Vortex	1	50	0.5	2.7	2.7	
			2	>4.4	>4.3	
		40	10			4-5

*1: Log (Initial) - Log (After treatment).
*2: *Alicyclobacillus* spp. spores.
Data presented by Ecolab Inc.

(b) Vortex™

Vortex™ (Ecolab) is a liquid sanitizing preparation mainly composed of the peroxy acids, peracetic acid and peroxycaprylic acid.

The sanitizing strength of peroxy acid depends on the length of its alkyl chain and, therefore, peroxycaprylic acid (C8) has stronger sanitizing power than peracetic acid (C2). This composition has a much better sanitization effect on spore-forming bacteria at a lower concentration than Oxonia Active, whose principal ingredient is peracetic acid. Therefore, Vortex™ is superior to the bacteriocide Oxonial Active (with peracetic acid as a unique peroxy acid), and works more efficiently on a wider range of microorganisms, especially on thermophilic acidophilic bacteria which are drug resistant [45)].

A sanitization effect of 4.3 D or more on *Alicyclobacillus* can be achieved by 50°C × 2 min contact of 1 % solution (Table 8-10). Vortex™ has a stronger effect against *Alicyclobacillus* spores than Oxonia Active, with a bactericidal effect of more than 4.3 D achieved in 2 minutes when contacting a 1 % solution of Vortex™, as opposed to 15 min for Oxonia Active. The pH of the Vortex™ stock solution does not exceed 1 and that of working solution is 2.5 at 1 % concentration, depending on the quality of the water [45].

(c) Oxine®

Oxine® (Nu-Calgon) is a bactericidal preparation whose principal component is chlorine dioxide. The preparation has a sanitizing strength of 99.998 % kill rate for *A. acidoterrestris* spores when contacted for 10 min at 30 ppm (Table 8-11). This sanitization effect is about 5 D [46].

Table 8-11. Efficacy of Oxine in killing spores of *A. acidoterrestris* [46]

Test Organisms	Concentration	Contact Time	Kill (%)
A. acidoterrestris spores	30 ppm	10 min	99.998
B. cereus spores	200 ppm	5 min	99.999

(4) Rapid cooling

Rapid cooling of pasteurized juice can help prevent spoilage by *Alicyclobacillus*. There are two possible explanations for the efficacy of this method: either spores stimulated to germinate by pasteurization are damaged by the rapid cooling and cannot continue to grow, or the spores are not regulated to grow at low temperatures.

Takashi Tanaka

8-3 Factors contributing to spoilage of food products, and spoilage prevention measures

8-3-1 Introduction

If *Alicyclobacillus* spores have not been eliminated during the production sanitization processes, it is very difficult to remove them from the final product. Countermeasures to prevent spores from germinating and growing in the final product include product design (vessel shape) and the use of antibiotics and preservatives. These are discussed below in further detail.

Keiichi Goto

8-3-2 Measures to prevent food spoilage

(1) Emulsifier

Sucrose fatty acid ester is an emulsifier that is effective for inhibiting the germination of *Alicyclobacillus* spores. It is used for retort-sterilized canned drinks such as coffee and milk drinks.

Sucrose palmitic acid ester (monoester-P) is a sucrose fatty acid ester, that has bacteriostatic and antibacterial effects against *Alicyclobacillus* spores (Table 8-12) [47]. Diglycerine monomyristate has equivalent preventive effects in drinks, as well as bacteristatic and antibacterial effects against *A. cycloheptanicus* (Table 8-13) [48]. The concentration of emulsifier additives must be increased to be effective in pulp-containing drinks like orange juice and apple juice, or fiber-containing vegetable juice drinks (Table 8-14) [47, 48].

Table 8-12. MIC values of monoester-P against *Alicyclobacillus* spores in clear juice drink [47]

Strain	Monoester-P (ppm)				
	0	10	25	50	75
A. acidoterrestris (DSM3923)	1×10^6	2×10^1	6×10^1	5×10^1	3×10^1
A. acidocaldarius (IFO15652)	2×10^6	3×10^2	5×10^2	4×10^2	4×10^2

Clear juice: pH 3.4, initial spores counts: 1×10^3 CFU/ml, after treatment (95 °C × 30 sec), storage (35 °C × 3 weeks).
Data presented by Mitsubishi-Kagaku Foods Corporation.

Table 8-13. MIC values of glycerin fatty acid ester against *A. cycloheptanicus* spores in near water [48]

Poemu DM-100	Counts after storage 30 °C (CFU/ml)	
(ppm)	3 days	7 days
0	2.8×10^4	7.6×10^5
2	4.2×10^1	4.6×10^1
3	2.9×10^1	3.4×10^1
5	3.3×10^1	4.0×10^1

Initial spores counts: 1.7×10^2 CFU/ml

Table 8-14. Effect of pulp in drink on growth inhibition of microorganisms by an emulsifier [49)]

Orange juice content (%)	Kind of juices	pH	Brix	Inhibitory concentration (ppm)
10	contained pulp	3.5	10	≧30
10	not contained pulp	3.5	10	≦10

Emulsifier: glycerin fatty acid ester.
Data presented by RIKEN VITAMIN Co. Ltd.

(2) Phytogenic antibacterial agents

(2)-1 Polyphenol

A. acidoterrestris does not grow well in red grape juice, reportedly due to the presence of high levels of polyphenols, which have antimicrobial activity (Table 8-15) [50)]. Some substances with antimicrobial activities have additive and synergistic antimicrobial activity. However, the minimum inhibition concentrations (MICs) of such substances range from 50-1000 μg/ml, which is much higher than concentrations found in grape juice (≤ 3 μg/ml). Therefore, *A. acidoterrestris* growth inhibition in red grape juice is not due simply to the synergistic effects of polyphenols. It should also be noted that polyphenols have stronger antimicrobial effects at lower pHs, lower effect in the presence of fibers (which adhere to polyphenols) [51)] and no effect in the higher pH range.

Table 8-15. Effect of grape polyphenols on the growth of *A. acidoterrestris* spores [50)]

Polyphenol	MIC (μg/ml)
Resveratrol	50
Ferulic acid	150
p-Coumalic acid	200
p-Hydroxybenzoic acid	400
(-)-Catechin-gallate	400
"Kyoho"proanthocyanidine	900
Caffeic acid	1000

(2)-2 Licorice oil extract

Sun Licorice is a commercial antimicrobial preparation whose effective ingredients are derived by oil extraction from licorice root. The antimicrobial ingredient is grablidin, which is contained at 1 % (w/v) concentration in Sun Licorice. The antimicrobial action of Sun Licorice against spores of *A. acidoterrestris* is effective in acidic drinks like juice drinks at 50-100 ppm (Table 8-16) [52].

Table 8-16. Inhibition of *A. acidoterrestris* growth in acidic drinks by liquorice extracts [52]

Acidic drinks	Inhibitory concentration of licorice extracts (ppm)								
	A. acidoterrestris strains								
	B2065	B2066	B2067	RB1	RB221	RB253	RB346	RB356	TAB-H1
Apple juice drink A	50	50	50	N.D.	50	N.D.	50	50	50
Lemon juice drink B	N.D.	100	N.D.	50	50	N.D.	100	50	N.D.
Grapefruit juice drink C	N.D.	150	100	100	100	N.D.	50	100	100
Grape (Red) juice drink D	50	50	50	N.D.	50	50	50	50	50
Grape (White) juice drink E	N.D.	50	N.D.	50	N.D.	50	50	N.D.	50
Juice drink F	50	100	50	50	50	N.D.	50	50	50
Near water A	N.D.	50	N.D.	N.D.	N.D.	N.D.	N.D.	50	N.D.
Near water B	50	50	50	50	50	50	50	50	50
Near water C	50	50	50	50	50	50	50	50	50
Sport drink	N.D.	N.D.	N.D.	N.D.	50	N.D.	N.D.	N.D.	50

Licorice extracts: SAIN LICORICE (containing 1 % glabridine, MITSUI NORIN CO.,LTD.).
N.D.: No data.

(2)-3 Antimicrobial peptide

Antimicrobial peptides called thionins are present in the endosperm of barley and wheat [53-57] and form 3 major groups: α-, β-, and γ-thionins. α- and β-thionins repress the growth of various kinds of molds [58], bacteria [58-60], and yeasts [61, 62] at concentrations in the range of $1\text{-}10^2$ μg/ml. The antimicrobial effects of γ-thionin are restricted to molds [63].

α- and β-thionins from barley and wheat inhibit growth of *A. acidoterrestris* ATCC 49025^T at 5-10 μg/ml concentration in YEPG (yeast peptone glucose broth) [64]. Wheat derived α-thionin at a concentration of 20 μg/ml effectively decreases the number of *A. acidoterrestris* ATCC 49025^T spores in mandarin orange juice or fruit-vegetable mix juice (Fig. 8-1) [65], but only has a weak effect in 30 % apple juice (Table 8-17). Moreover, a 0.2 M citric acid extract of barley seeds (α-thionin; 100 μg/ml, β-thionin; 50 μg/ml) and 3 % (v/v) malic acid inhibits the growth of *A. acidoterrestris* ATCC 49025^T (Table 8-17) [64] in mandarin orange juice.

Takashi Tanaka

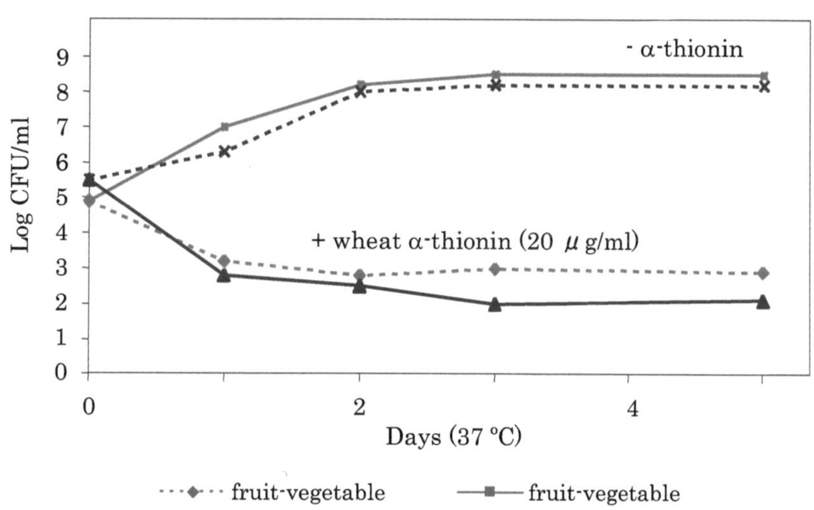

Fig. 8-1. Growth inhibition of *A. acidoterrestris* by a-purothionin in fruit juice [65]

A. acidoterrestris spores were inoculated into two juice types, and then growth was monitored for five days in the presence and absence of wheat a-thionin.

Table 8-17. Growth inhibition of *A. acidoterrestris* spores by thionins [64)]

Tinonin	Amount of addition	Juice	Initial spores counts (CFU/ml)	Counts after storage 37 °C (CFU/ml)	
				1 days	5 days
		100 % satsuma mandarin	3.0×10^5	6.3×10^2	1.3×10^2
α-thionins from wheat	20 μg/ml	fruit-vegetable mixture	7.9×10^4	1.6×10^3	7.9×10^2
		30 % dilution of apple	5.0×10^4	1.6×10^3	1.9×10^4
Barley-citric acid extract*	3 %	100 % satsuma mandarin	2.0×10^5	nt	$<1.0 \times 10^3$
Barley-malic acid extract	3 %	100 % satsuma mandarin	2.0×10^5	nt	$<1.0 \times 10^3$

*: Containing α-Thionin 100 μg/ml and β-thionin 50 μg/ml.
nt: Not tested.

(2)-4 Acerola fruit juice

Acerola fruit (*Malpighia emarginata* DC.) has been reported to have growth inhibitory activity against *Alicyclobacillus acidoterrestris* [80, 81]. The growth of *A. acidoterrestris* ATCC 49025^T in YSG medium was completely inhibited when a water-extracted powder of acerola fruit (Vietnamese cultivar) was added at the concentration of 0.2 % (w/v) (Fig. 8-2). This powder also inhibited the growth of *A. acidoterrestris* ATCC 49025^T in 35 % (w/v) apple juice at the concentration of 0.1 % (w/v) (Fig. 8-3). Furthermore, it has been reported to have inhibitory effects on other TAB species.

Acerola fruit is characterized by an extremely high content of vitamin C. It also contains glucose, fructose, and malic acid as its major components, and a certain amount of polyphenols. Partially-purified substances of these acerola components (malic acid, vitamin C and polyphenols) did not show inhibitory activities comparable to that of intact acerola fruit extract when used at the corresponding concentrations. Thus, it is assumed that the inhibitory activity of acerola fruit originates from the combined effect of those major and minor components.

Acerola fruit juice could be a useful nutritious ingredient that has the efficacy to prevent deterioration of fruit beverages by TAB.

Shozo Tanada

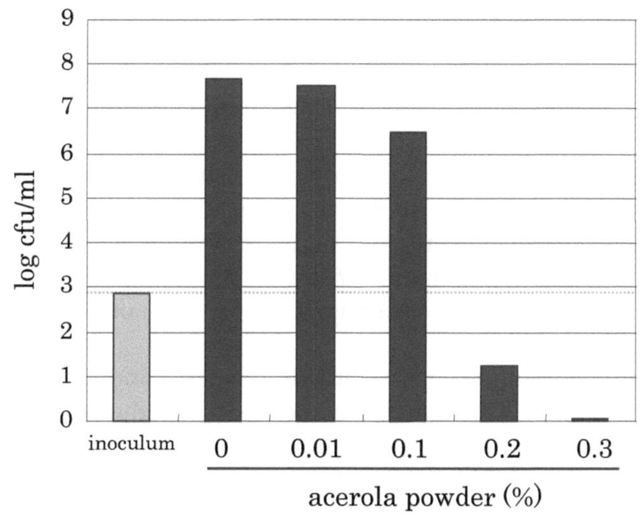

Fig. 8-2. Growth inhibition of *A. acidoterrestris* ATCC 49025^T in YSG medium by acerola fruit powder [80, 81]

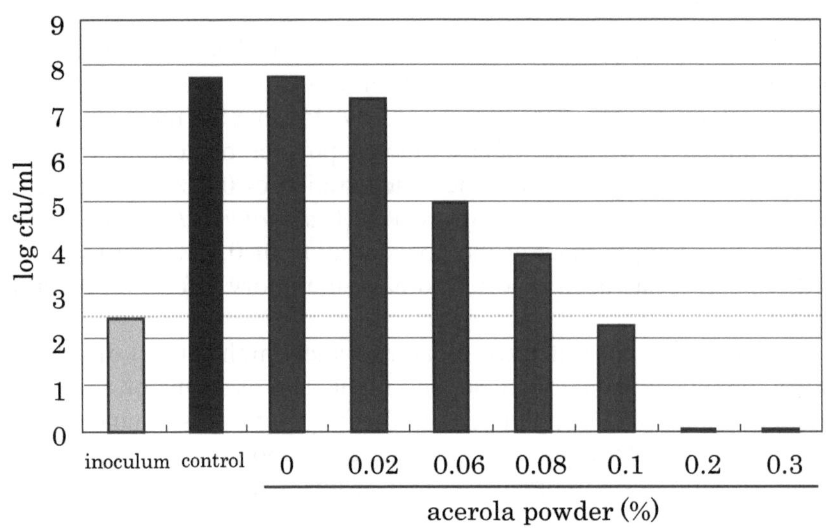

Fig. 8-3. Growth inhibition of *A. acidoterrestris* ATCC 49025^{T} in 35 % apple juice by acerola fruit powder [80, 81]

(3) Preservatives

(3)-1 Nisin

Nisin, derived from *Lactococcus lactis*, is the first bacteriocin to gain recognition as "generally regarded as safe" (GRAS) status from the FDA (U.S. Food and Drug Administration). It is currently used as a food preservative in 56 countries, mainly in Europe. In Japan, nisin has not yet been designated as approved for use as a food additive, but this is currently under review. Nisin exerts a high antibacterial activity against gram-positive bacteria including *Listeria monocytogenes*, *Clostridium botulinum*, and other food-poisoning bacteria, but not in Gram-negative bacteria, molds, and yeasts [66, 67].

Nisin is also known to suppress growth of spores and vegetative cells of *A. acidoterrestris* [68, 69]. In addition, a bacteriocin produced by *Lactococcus* sp. CU216 shows inhibitory activity against spores and vegetative cells of *Alicyclobacillus* [70]. The MIC of nisin for spores ranges from less than 1 to 13 IU/ml on modified yeast peptone glucose agar medium (mYPGA) at pH 3.4, and from 25 to 100 IU/ml at pH 4.2. A lower pH results in a lower MIC value. The MIC value of nisin for vegetative cells is higher than that for spores (Table 8-18). Although growth is suppressed in orange fruit mix juice by the addition of 25-50 IU/m nisin, it is not inhibited in the clear apple juice, even by the addition of nisin at more than 600 IU/ml.

Table 8-18. MIC values of nisin against vegetative cells and spores of *A. acidoterrestris* on mYPGA plates [68)]

Strains	MIC (IU/ml)			
	mYPGA (pH 3.4)*		mYPGA (pH 4.2)	
	Cells	Spores	Cells	Spores
ATCC 49025^T	25	<1	50	25
ATCC 49026	25	2	50	25
ATCC 49027	25	13	50	25
AB-1	3	<1	100	50
AB-2	2	<1	25	25
AB-4	2	2	25	25
AB-5	50	13	100	100

*: modified Yeast-Peptone-Glucose Agar.

(3)-2 Lysozyme and NaCl

The growth of *A. acidoterrestris* is inhibited in Orange Serum Broth (OSB) by adding lysozyme (a food preservative), at a concentration of 0.001 % (w/v) and likewise, by adding sodium chloride (NaCl) at a concentration of 5 % (w/v) (Table 8-19) [71)].

Table 8-19. Effect of lysozyme and NaCl on growth of *Alicyclobacillus* [71)]

Strains	0.001 % lysozyme	5 % NaCl
A. acidoterrestris N-1089 (spoiled tomato juice)	-	-
A. acidoterrestris N-1090 (spoiled mix juice)	-	-
A. acidoterrestris N-1098 (conc. apple juice)	-	-
A. acidoterrestris N-1104 (apple pear juice)	-	-
A. acidoterrestris N-1107 (spoiled apple juice)	-	-
A. acidoterrestris N-1108 (apple-cranberry juice)	-	-
A. acidoterrestris ATCC 49025^T	-	-
A. acidocaldarius ATCC 27009^T	+	+
A. cycloheptanicus ATCC 49028^T	-	-

Orange Serum Broth, +: positive, -: negative.

(3)-3 Organic acids

Organic acids are sour compounds found naturally in abundance in fermented food and fruits, and are known to have antibacterial activities.

The growth of *A. acidocaldarius* AC-1 strain derived from a fruit juice drink is suppressed by organic acids at pH 5.5 or below. Stronger effects are observed for acetic acid, adipic acid, lactic acid, and fumaric acid. In particular, acetic acid has very strong growth inhibitory effects [72]. Antibacterial activities rely on the undissociated molecules of organic acids, and dissociation is more strongly repressed at lower pH. Since acetic acid has the lowest acid dissociation constant, it has the strongest antibacterial activity of these organic acids.

Since *Alicyclobacillus* is an obligately aerobic bacterium, availability of dissolved oxygen has a great influence on its growth ability. Ascorbic acid completely inhibits the growth of *A. acidoterrestris* when added to apple juice at a concentration of 0.15 mg/ml [73].

(3)-4 Ethanol

Ethanol has a bacteriocidal effect and it is used as a sanitizer. Although it has no bacteriocidal effect at low concentration (<10 %), it shows antibacterial activity in general. Growth of *Alicyclobacillus* is suppressed by ethanol at concentrations of ≥6 % (v/v) [74].

(4) Containers

Because *Alicyclobacillus* is an obligately aerobic bacterium, its growth activity is affected by the availability of dissolved oxygen. Dissolved oxygen levels are determined by the oxygen concentration in the headspace of the vessel. In the case of long-life paper pack (produced by TetraPak), there is no head space. If there is no way to remove the head space, it is replaced with nitrogen gas.

The ability of oxygen gas to pass through the packaging container material also influences the growth of *A. acidoterrestris. A. acidoterrestris* grows most easily in PET bottles which have moderate oxygen transparency, followed by LL paper packs, and then cans which do not allow any air transfer. From outside to inside, the LL paper pack produced by Tetra Pak consists of six layers of polyethylene, paper, polyethylene, aluminum sheet, second layer polyethylene, and innermost polyethylene [75].

PET bottles are the most widely used kind of packaging in Japan, and this is thought to be one of the reasons for the high incidence of contamination due to *Alicyclobacillus*. In recent years, as the demand for PET bottles has increased in Europe there is a danger that incidents of soft drink contamination may increase due not only to *Alicyclobacillus* but also to other aerobic spore-forming bacteria.

(5) Storage temperature

The growth temperature range for *Alicyclobacillus* is between 20 and 70°C, depending on the species. For *A. acidoterrestris*, which is the main cause of off-odors in fruit juice drinks, the growth temperature range is from 20 to 55°C [76]. Determination of the growth temperature range requires several weeks, as the growth rate is very slow at extremes of the temperature range. [77].

It is expected that spoilage by *A. acidoterrestris* can mostly be minimized by maintaining products at 20°C or below during distribution.

Takashi Tanaka

8-3-3 Influence of product design on the risk of contamination by *Alicyclobacillus*

There is extensive literature describing the influence of product design on the risk of microbial contamination of soft drinks [78]. The first step in designing a product is to assess the risk of microbial contamination of the foodstuff. The parameters listed below may be included in the assessment; note that this list is not exhaustive, and assessment criteria vary among companies.

(a) No microbial growth is observed, but an unpleasant (off) taste or odor, or cloudiness is observed in the product.

(b) Microbial growth is allowed under the conditions in (a), to the extent that cloudiness is not observed (less than 10^6 cells/ml). Care should be taken in this case though, since an unpleasant-odor may be present even if no muddiness is observed.

(c) A cloudy appearance or unpleasant-odor is permissible if it is masked by palatable characteristics of the product.

(d) A certain level of contamination is permissible if the distribution temperature and conditions are adhered to. The risk of contamination will be low if the distribution temperature and time are enforced. Guidelines and evaluation methods, and supplemental information specific to *Alicyclobacillus* are given below.

(1) Guidelines for Alicyclobacillus

Guidelines to minimize *Alicyclobacillus* contamination of food products have not been defined either internationally or domestically. Both IFU and Japanese producers have identified that *A. acidoterrestris* requires the closest monitoring. However, other *Alicyclobacillus* species, some of which produce guaiacol, also cause food spoilage and must be monitored. Methods to detect microorganism contaminants are difficult to prescribe as they must be sufficiently sensitive to detect as little as a single spore, which has been reported to cause spoilage of 500 ml of product [79]. Local guidelines tend to be based on the results of inoculation and sanitization tests (see below), since microbial growth varies according to the type of juice and the bacterial species and strain used as inoculum. The International Society of Beverage Technologists is now working towards setting guidelines for *Alicyclobacillus*.

(2) Evaluating the growth of Alicyclobacillus in food products

(2)-1 Inoculation test

The inoculation test involves inoculating spores (or vegetative cells) into food products, and then monitoring bacterial growth and the release of unpalatable (off) odors. The test provides an indication of the risk of contamination by *Alicyclobacillus*, allowing appropriate precautionary measures to be implemented during production, packaging and storage of the product.

Reliable results are obtained when multiple strains are used for the inoculation test. However, in practice, usually only two strains are assessed and cell (spore) counts are between 10 and 10^3 per product. Criteria for the selection of strains and methods for the preparation of spore inoculum are described later.

After inoculation, the sample is heat shocked by treatment at 80°C × 10 min or 70°C × 20 min, with cooling immediately to cultivation temperature after treatment. Alternatively, pasteurization conditions are used in some cases. The heat shock treatment is omitted when vegetative cells are used as the inoculum. Cultivation is then in the range of 25 to 45°C. Depending on the purpose of the test, cultivation conditions are determined by the distribution temperature and the optimal growth temperatures of the strains used as inoculum. Several incubation periods are usually assessed, ranging in duration from several days up to the product expiry date, or longer (more than half a year).

Cell growth and the release of unpleasant odors are then assessed. Agar media or membrane filters are generally used to examine cell growth, (see

Chapter 4), and gas chromatography is used to characterize the source of unpleasant odors (see Chapter 5). In recent years, the peroxidase method has been developed to detect guaiacol (see Chapter 5).

Characteristics intrinsic to the product may obscure test results. Specifically, if guaiacol precursors are absent from the product, *A. acidoterrestris* will not grow to a detectable level. Further, oxygen availability (open versus unopened containers) and product container design will influence microbial growth. These factors must be taken into consideration when interpreting test results.

Details concerning factors that become parameters are explained in Chapters 3 and 4.

(2)-2 Sanitization test

The sanitization test involves inoculation of cells (usually spores) into the test product, followed by sanitization via prescribed conditions (heating, preservatives, chemicals etc.). The number of surviving cells is then determined. Variables such as temperature and inoculum cell count must be combined in all possible combinations in order to determine the D or Z value, which gives an indication of sanitization efficacy.

(2)-3 Test example 1

Vegetative cells of three *A. acidoterrestris* strains (derived from raw materials) were inoculated into two kinds of juice-containing soft drinks (hot-pack-filled PET bottle) with different formulations and incubated at 40°C. Viable cells were counted at prescribed intervals. In one drink formulation, only one of the three *A. acidoterrestris* strains grew, while in the other formulation none of the strains grew. Thus the drink formulation which did not support *A. acidoterrestris* growth was selected as the final product formulation.

(2)-4 Test example 2

Spores of three *A. acidoterrestris* strains derived from bulk juices were inoculated into a juice-containing soft drink. After HTST sanitization, the soft drink was hot-packed into PET bottles and stored at 40°C. Viable cells were counted and off-odor components were evaluated as above. Although cell counts increased by only 2-3 times, HTST sanitization conditions were reevaluated because (i) the off-odor component (guaiacol) exceeded the acceptable threshold and (ii) in the sensory test, an unpleasant odor which differed from that of the negative control sample was detected.

(2)-5 Test example 3

Spores of eight strains of *A. acidoterrestris* (derived from various materials) were inoculated into a sports drink (commercial PET bottle product) with the addition of a preservative up to a final concentration in the range of 1-100 ppm. The PET bottles were stored at 35°C and viable cells were counted at prescribed intervals after treatment at 70°C for 20 min. From the results, the MIC value of the preservative was determined to be 10-25 ppm. Consequently, it was concluded that for this product, contamination by *A. acidoterrestris* could be prevented by adding the preservative at a concentration of 25 ppm.

(Note 1) *A. acidoterrestris* strains

Test results are influenced by the *A. acidoterrestris* strain type and the type of food being tested (see 4-2-6). Therefore, it is recommended to use multiple strains derived from raw materials, or strains derived from same raw materials as inoculation medium. However, when such strains cannot be isolated, type strains or standard strains are acceptable substitutes. Type strains and standard strains are available domestically from NBRC (http://www.nbrc.nite.go.jp/) or JCM (http://www.jcm.riken.go.jp/), or internationally from DSMZ (http://www.dsmz.de) or ATCC (http://www.atcc.org). Instructions for re-hydration are provided by each culture collection. About ten *Alicyclobacillus* strains derived from drink products and their materials are kept in the IAM culture collection.

(Note 2) Cell suspension

For the preparation of a spore suspension, the bacterial strain of interest is spread on YSG agar media, then incubated at the appropriate temperature for 1-10 days. An optical microscope is used to determine when >90 % of the cells are spores. Cells are collected from the agar surface using sterilized buffer (0.1 % $MgSO_4 \cdot 7H_2O$, 0.8 % NaCl: pH 4.5), centrifuged (ca. 3500 rpm × 10 min) and then washed several times with the same buffer. The final cell preparation is suspended in sterilized distilled water and stored at 4°C for up to 30 days. When a stored spore suspension is used as inoculum, it is desirable to determine the number of viable spores prior to inoculation. Various media types should be tested for each strain to determine which type best supports sporulation. Sporulation rates of isolated strains are generally higher than those of type strains, and sporulation rates tend to be higher on solid agar rather than in liquid media.

Vegetative cells are often weakened or damaged during the gathering and washing processes. Consequently, it is often very difficult to obtain consistent results in replicate tests. The consistency of results is improved when vegetative cells are pre-incubated in a liquid medium or in the product itself, however, the effects of foreign matters and metabolites should be then be taken into consideration.

Tetsuya Sawaki, Rie Yamamoto, Keiichi Goto

8-4 Spoilage factors related to distribution and prevention measures

For products that have been contaminated by *Alicycobacillus* spores, environmental temperature during storage and distribution is the most important factor to consider. Spoilage does not occur when products are stored at ≤ 20°C as explained before (see Chapter 6). However, when products are stored and distributed at ambient temperatures, extremes as high as 30-40°C may be reached, for example in shop fronts, warehouses and delivery trucks. If all products could be distributed and stored at ≤ 20°C, *Alicyclobacillus* contamination would not occur, but unfortunately in practice this is very difficult to achieve.

Japan's climate is classified as "humid subtropical with a hot summer" (Cfa by Köppen's classification) except for Hokkaido whose climate is "Marine west coast with warm summer" (Cfb) and "humid continental with warm summer" (Dfb). Therefore, it exceeds 35°C during summer days. Such climatic conditions are likely factors that contribute to the occurrence of *Alicyclobacillus* contamination incidents.

Keiichi Goto

References

1. Uchino F, Doi S (1967) Acido-thermophilic bacteria from thermal waters. Agric Bio Chem 31:817-822.
2. Darland G, Brock TD (1971) *Bacillus acidocaldarius* sp. nov., an acidophilic thermophilic spore-forming bacterium. J Gen Microbiol 67:9-15.
3. Deinhard G, Saar J, Krischke W, Poralla K (1987b) *Bacillus cycloheptanicus* sp. nov., a new thermoacidophile containing ω-cycloheptanes fatty acids. Syst Appl Microbiol 10:68-73.
4. Goto K, Tanimoto Y, Tamura T, Mochida K, Arai D, Asahara M, Suzuki M, Tanaka H, Inagaki K (2002b) Identification of thermoacidophilic bacteria and a new *Alicyclobacillus* genomic species isolated from acidic environments in Japan. Extremophiles 6:333-340.

5. Cerny G, Hennlich W, Poralla K (1984) Fructsaftverderb durch Bacillen: isolierung und charakterisierung des verderbserregers. Z Lebens Unters Forsch 179:224-227 (In German).
6. Deinhard G, Blanz P, Poralla K, Altan E (1987) *Bacillus acidoterrestris* sp. nov., a new thermotolerant acidophile isolated from different soils. Syst Appl Microbiol 10:47-53.
7. Sprittstoesser DF, Churey JJ, Lee CY (1994) Growth characteristics of aciduric sporeforming bacilli isolated from fruit juices. J Food Protection 57:1080-1083.
8. Wisse CA, Parish M (1998) Isolation and enumeration of sporeforming, thermo-acidophilic, rod-shaped bacteria from citrus processing environments. Dairy Food Environ Sanitation 18:504-509.
9. Komitopoulou E, Boziaris IS, Davies EA, Delves-Broughton J, Adams MR (1999) *Alicyclobacillus acidoterrestris* in fruit juices and its control by nisin. Int J Food Sci Tech 34:81-85.
10. Walls I, Chuyate R (2000) Isolation of *Alicyclobacillus acidoterrestris*. J AOAC Int 83:1115-1120.
11. Walls I, Chuyate R (2000) Spoilage of fruit juices by *Alicyclobacillus acidoterrestris*. Food Australia 52:286-288.
12. Jensen N (2000) *Alicyclobacillus* in Australia. Food Australia 52:282-285.
13. Eguchi SY, Manfio GP, Pinhatti MEC, Azuma EH, Variane S. Colecao de Culturas Tropical (CCT) Fundacae Andre Tossello and ABECitrus (1999) Detection method, ecology and involvement in the deterioration of fruit juices study.
14. Duong HA, Jensen N (2000) Spoilage of iced tea by *Alicyclobacillus*. Food Australia 52:292.
15. Goto K, Matsubara H, Mochida K, Matsumura T, Hara Y, Niwa M, Yamasato K (2002) *Alicyclobacillus herbarius* sp. nov., a novel bacterium containing ω-cycloheptane fatty acids, isolated from herbal tea. Int J Syst Evol Microbiol 52:109-113.
16. DeLucca II AJ, Kichen RA, Clarke MA, Goynes WR (1992) Mesophilic and thermophilic bacteria in a cane sugar refinery. Zuckerind 117:237-240.
17. Yoshino M (2002) European circumstances of thermotolerant, acidophilic bacteria. Report of Japan Fruit Juice Association 10:23-30 (in Japanese).
18. Eguchi SY, Manfio GP, Pinhatti MEC, Azuma EH, Variane S. Colecao de Culturas Tropical (CCT) Fundacao Andre Tosello and ABECitrus (1999) Detection methods, ecology and involvement in the deterioration of fruit juice study.
19. Goncarves AC (2003) "Comprehension of tab research of abecitrus and tab control now and in the future", Proceeding of the ILSI Japan Symposium on Thermotolerant, Acidophilic Bacteria.
20. Goto K (2000) Thermoacidophilic spore-forming bacteria: *Alicyclobacillus*. J Antibacterial Antifun Agents Japan 28:13-38 (in Japanese).
21. Eguchi SY, Manfio GP, Pinhatti ME, Azuma E, Variane SF (1999) Acidothermophillic Sporeforming Bacteria (ATSB) in Orange Juices: Detection

Methods, Ecology, and Involvement in the Detection of Fruit Juices. Report of the Research Project, Brazil.
22. Goto K (2003) Fundamentals of Thermotolerant acidophilic bacteria and prevention methods for them. Abstracts of the ILSI JAPAN Symposium on Thermotolerant Acidophilic Bacteria, pp. 12-18 (in Japanese).
23. Lee S, Dougherty RH, Kang D (2002) Inhibitory effects of high pressure and heat on *Alicyclobacillus acidoterrestris* spores in apple juice. Appl Environ Microbiol 68:4158-4161.
24. Shearer AEH, Dunne CP, Sikes A, Hoover DG (2000) Bacterial spore inhibition and inactivation in foods by pressure, chemical preservatives, and mild heat. J. Food Protection 63:1503-1510.
25. Chuma M, Todoroki S (2003) Radiosensitivity of the thermotolerant acidophilic bacterium *Alicyclobacillus acidoterrestris*. Abstracts of the 40th meeting of the radioisotope and radiation studies on science and technology, p. 131 (in Japanese).
26. Ikegami Y, Enda A, Matsui T, Nakagawa K (2003) Characteristics of the acidophilic bacteria isolated from the fruit juice beverage. Report of Toyo Junior College of Food Technology and Toyo Institute of Food Technology 21:105-115 (in Japanese).
27. Yamazaki K, Isoda C, Tezuka H, Kawai Y, Shinano H (1997) Thermotolerence and control of the acidic drink-spoiling bacterium *Alicyclobacillus acidoterrestris*. Journal of the Japanese Society for Food Science and Technology 44:905-911 (in Japanese).
28. Hayasaka I (2003) Application of the filter effective for elimination of the spores of thermotolerant bacteria. Monthly Food Chemical 9:9-12 (in Japanese).
29. Matsumoto K (2001) Elimination of spores of the thermotolerant bacteria. Food and Development 36:10-11 (in Japanese).
30. Matsuda T (1985) Chemicals and microbial control. Handbook of Application of Sanitation and Elimination, p. 192, Science Forum (in Japanese).
31. Akabane Y (1991) Microbial control effect by alcohol and supplements. Current data collection of food microbial control systems, p. 464, Science Forum (in Japanese).
32. Shibasaki I (1983) Chemical sanitation. New Food Sanitation Technology, pp. 241, Korin (in Japanese).
33. Nakayama M, Fujimoto A, Higuchi A, Watanabe M, Iio M, Miyamoto T (2003) The effect and character of slightly acidified hypochlorous acid water on the *Bacillus* spores and *Lactococcus*. Journal of Antibacterial and Antifungal Agents Japan. 31:421-425 (in Japanese).
34. Shibasaki I (1983) Chemical sanitation. New Food Sanitation Technology, pp 249, Korin (in Japanese).
35. Shibasaki I (1980) Sanitation of microbes in aseptic packaging. Packaging Systems and Hygiene, Vol. 4, p. 55, Science Forum (in Japanese).
36. Nozawa E (1997) The vessel sanitation technology on the aseptic filling packaging. The Soft Drinks Technology 2:65-73 (in Japanese).

37. Tanabe T (1996) Peracetic acid sanitizers used on aseptic filling systems. Beverage Japan 19:51-56 (in Japanese).
38. Ishida T, Yoneuchi S (1986) New Technology of Ozone Utilization, p. 386, Sanshu Shobo. (in Japanese).
39. Tsuji S (1984) Cleaning and Sanitation in Food Factories, p. 171, Kenpakusha.
40. Orr RV, Beuchat LR (2000) Efficiency of disinfectants in killing spores of *Alicyclobacillus acidoterrestris* and performance of media for supporting colony development by survivors. J Food Protection 63:1117-1122.
41. Eguchi R (1991) A compilation of sanitizers. Sanitizer Handbook, Nissoken Publishing (in Japanese).
42. Gray PM, Kang DH, Dougherty RH (2002) Controlling *Alicyclobacillus acidoterrestris* spores with chlorine dioxide in laboratory medium and apples. Session 15D, Food Microbiology: Fruits and vegetables, Annual Meeting and Food Expo - Anaheim, California.
43. Goncalves AC (2003) Incidents and Control of TAB in fruit juices in Brazil. Abstracts of the ILSI JAPAN symposium on the Thermophilic Acidophilic Bacteria, pp. 19-27.
44. Cerf O, Hermier J (1972) Le Lait 52:1.
45. Materials for Vortexx™, Ecolab (in Japanese).
46. Oxine® flyer, http://www.nucalgon.com/.
47. Materials for Ryoto® Monoester-P, "Suppression effect of sugar fatty acid ester on the acidophilic spore-forming bacteria in drinks." Mitsubishi Chemical Foods (in Japanese).
48. Nakanishi H (1999) Utilization of a emulsifier for drinks. Beverage Japan 22:93-95 (in Japanese).
49. Diglycerine monomyristate data, "Growth of thermotolerant acidophilic bacteria in acidic drinks and bacteriostat effect on them." Riken Vitamin (in Japanese).
50. Oita S, Kamiyama N (2002) Antibacterial effect on thermoacidotolerant bacterium *Alicyclobacillus acidoterrestri*s. Journal of the Japanese Society of Food Science and Technology 49:555-558 (in Japanese).
51. Niwa M (2001) Lecture of Classification and Identification of Microorganisms: Classification and characterization of newly found thermotolerant acidophilic bacteria. Soft Drink Technology, pp. 9-26 (in Japanese).
52. Hojo H (2003) Antibiotic activities of the licorice extract preparation, Sun Licorice. Monthly Food Chemical 6:116-122 (in Japanese).
53. Ozaki Y, Wada K, Hase T, Matsubara H, Nakanishi T, Yoshizumi H (1980) Amino acid sequence of purothionin homolog from barley flour. J Biochem (Tokyo) 87:549-555.
54. Wada K (1982) Localization of purothionins and genome expression in wheat plants. Plant Cell Phyciol 23:1357-1361.
55. Mendez E, Moreno A, Colilla F, Pelaez F, Limas GG, Mendez R, Soliano F, Salinas M, de Haro C (1990) Primary structure and inhibition of protein synthesis in eukaryotic cell-free system of novel thionin, gamma-hordothionin, from barley endosperm. Eur J Biochem 194:533-539.

56. Hase T, Matsubara H, Yoshizumi H (1978) Disulfide bonds of purothionin, a lethal toxin for yeasts. J Biochem (Tokyo) 83:1671-1678.
57. Collila FJ, Rocher A, Mendez E (1990) γ-Purothionins: amino acid sequence of two polypeptides of new family of thionins from wheat endosperm. FEBS Lett 270:191-194.
58. Molina A, Ahl GP, Feaile A, Sanchez-Monge R, Garcia-Olmed G (1993) Inhibition of bacterial and fungal plant pathogens by thionins of types I and II. Plant Sci 92:168-177.
59. Okada T, Yoshizumi H, Terashita Y (1970) A lethal toxic substance for brewing yeast in wheat and barley. Part I. Assay of toxicity on various grains, and sensitivity of various yeast strains. Agric Biol Chem 34:1084-1088.
60. Okada T, Yoshimizu H (1973) Lethal toxic substance for brewing yeast in wheat and barley. Part III. Mode of action of toxic protein in wheat and barley on brewing yeast. Agric Biol Chem 37:2289-2294.
61. Fernandez de Caleya R, Gonzalez-Pascul B, Garcia-Olmedo F, Carbonero P (1972) Susceptibility of phytopathogenic bacteria to wheat purothionins *in vitro*. Appl Microbiol 23:998-1000.
62. Florack DEA, Visser B, de Vries PM, van Vuurde JWA, Stiekema WJ (1993) Analysis of the toxicity of purothionins and hordothionins for plant pathogenic bacteria. Neth J Plant Pathol 99:259-268.
63. Broekaert WF, Terras FRG, Cammue BPA, Osborn RW (1995) Plant defensins: novel antimicrobial peptides as components of the host defense system. Plant Physiol 108:1353-1358.
64. Oita S (2002) Control of thermoacidophilic *Alicyclobacillus acidoterrestris* by barley and wheat α- and β-thionins. Bull Natl Agric Res Cent West Reg 1:49-59.
65. Oita S (2002) Bactericidal effect of α-thionin from barley and wheat. Inf Natl Agric Res Cent West Reg 2001:325-326 (in Japanese).
66. Nettles CG, Barefoot SF (1993) Biochemical and genetic characteristics of bacteriocins of food-associated lactic acid bacteria. J Food Prot 56:338.
67. Eijsink VGH, Skeie M, Middelhoven PH, Brurberg MB, Nes IF (1998) Comparative studies of class II a bacteriocins of lactic acid bacteria. Appl Environ Microbiol 64:3275-3281.
68. Yamazaki K, Murakami M, Kawai Y, Inoue N, Matsuda T (2000) Use of nisin for inhibition of *Alicyclobacillus acidoterrestris* in acidic drinks. Food Microbiol 17:315-320.
69. Komitopoulou E, Boziaris IS, Daves EA, Delves-Broughton J, Adams MR (1999) *Alicyclobacillus acidoterrestris* in fruit juices and its control by nisin. Int J Food Sci Tech 34:81-85.
70. Oh S, Churey JJ, Worobo RW (1999) Inhibitory activity of *Alicyclobacillus* strains by bacteriocin of *Lactococcus* sp. CU216. IFT Annual Meeting, 37D-32.
71. Walls I, Chuyate R (1998) *Alicyclobacillus*-Historical perspective and preliminary characterization study. Dairy, Food Environ. Sanitation 18:499-503.
72. Ikegami Y, Enda A, Matsui T, Nakagawa K (1996) Characteristics of the acidophilic bacterium isolated from a fruit juice beverage. Rep. Toyo Junior

Collage of Food Technology and Toyo Institute of Food Technology (1994-1995) 21:105-115 (in Japanese).
73. Cerny G, Duong HA, Hennlich W, Miller S (2000) *Alicyclobacillus acidoterrestris*: influence of oxygen content on growth in fruit juices. Food Australia 52:289-291.
74. Splittstoesser DF, Lee CY, Churey JJ (1998) Control of *Alicyclobacillus* in the juice industry. Dairy Food Environ. Sanitation 18:585-587.
75. Nakamura M (2004) Problems on the quality control of aseptic filling packaging in paper vessels. Abstracts of the seminar on the aseptic packaging system 2004, pp. 7-8, Science Forum (in Japanese).
76. Goto K (2003) The newest fundamental knowledge of the thermotolerant, acidophilic bacteria and countermeasures against them. Proceedings of the ILSI Japan symposium on the thermotolerant, acidophilic bacteria, ILSI Japan (in Japanese).
77. Goto K (2000) Thermoacidophilic endospore-forming bacteria: *Alicyclobacillus*. Journal of Antibacterial and Antifungal Agents Japan. 28:499-508 (in Japanese).
78. Sawaki T (2003) Points to consider in microbial control of soft drinks. Beverage Japan 253:43-46 (in Japanese).
79. Yoneda H (2003) Inspection methods for thermotolerant, acidophilic bacteria. Report of Japan Fruit Juice Association 11:1-8 (in Japanese).
80. Tanada S, Tajima R, Koizumi T, Yamamoto S, Hanamura T, Mayama C, Aoki H (2006) Antimicrobial activity of Acerola fruit juice against Thermo Acidophilic Bacilli.Japanese Society of Food Microbiology, 27 th annual review C05 (in Japanese).
81. Tanada S, Tajima R, Koizumi T, Yamamoto S, Hanamura T, Mayama C, Aoki H (2006) Antimicrobial activity of Acerola fruit juice against *Alicyclobacillus acidoterrestris* in various kinds of Fruit juice. Japanese Society of Food Microbiology, 27 th annual review C06 (in Japanese).

Chapter 9: Agencies, associations, NPOs, institutes and researchers involved with *Alicyclobacillus*

This chapter introduces public organizations that conduct investigations and research on *Alicyclobacillus*, in various regions and countries, and indicates their research interests and work. Addresses of contact personnel, however, are not stated.

9-1 Japan

9-1-1 Japan Fruit Juice Association

Homepage address: http://www.kaju-kyo.ecnet.jp/

In November, 2002, the Japan Fruit Juice Association formed "a working group to designate a unified detection method for thermo tolerant, acidophilic bacteria" under the Juice Technical Research Committee and nine soft drink makers. With the cooperation of researchers from member companies, this detection method was established and published in Japan (Kaju Kyokai Hou, No. 535, March 25, 2004). The method was announced at the meeting of the microbiology working group of the International Federation of Fruit Juice Producers (IFU) in April, 2004 and May, 2005 (held in Germany), and is an active step towards international harmonization.

9-1-2 International Life Science Institute Japan (ILSI Japan)

Homepage address: http://www.ilsijapan.org/ilsijapan.htm

The Microbiology Subcommittee of the Food Safety Committee ILSI Japan, conducted a survey to determine which microorganisms were problematic in member companies of ILSI Japan, and from this information, determined target microorganisms (*Alicyclobacillus*) and published a handbook on the thermophilic acidophilic bacteria. In August, 2003 ILSI Japan held an international symposium, inviting such speakers as Mr. Bob Hartog from IFU and Mr. Antonio Carlos Gonçalves from ABECitrus to give insight into the issues regarding *Alicyclobacillus*.

9-2 Brazil

9-2-1 The Brazilian Association of Citrus Exporters (ABECitrus)

Homepage address: http://www.abecitrus.com.br/

ABECitrus, located in São Paulo State, is an association of Brazilian citrus exporting companies,. Their technical committee has, as already mentioned in this book, taken a leading role in conducting comprehensive research into *Alicycobacillus* (The report can be downloaded from ABECitrus homepage). Moreover, ABECitrus has been conducting fundamental research and studies regarding the control and inspection methods for *Alicycobacillus* with various organizations.

9-2-2 State University of Campinas (Universidade Estadual de Campinas)

Homepage address: http://www.unicamp.br/

The State University of Campinas is located in Campinas City of State São Paulo. Dr. Pilar Rodriguez de Massaguer, Food Engineering Course, has been carrying out research on *Alicyclobacillus* (identification, control measures, detection methods, and so on) in collaboration with ABECitrus and CCT.

9-2-3 Instituto de Tecnologia de Alimentos (ITAL)

Homepage address: http://www.ital.sp.gov.br/

ITAL is a food technological institute located in Campinas City, São Paulo. They perform various kinds of testing on demand, and are also conducting studies on *Alicyclobacillus*.

9-2-4 Coleção de Culturas Tropical (CCT), Fundação André Tosello

Homepage address: www.cct.org.br/

CCT is one of the world's most eminent microbiology institutes, located in Campinas City in Brazil, and has carried out comprehensive studies on *Alicyclobacillus* in collaboration with ABECitrus. Their research has been directed towards establishing a method for detection of *Alicyclobacillus* and establishing quality control indices, by inspecting, for example, the diversity of isolated microorganisms at the gene level.

9-3 Europe

9-3-1 International Federation of Fruit Juice Producers (IFU)

Homepage address: http://www.ifu-fruitjuice.com/

IFU is an association of fruit juice producers, with companies from 33 countries as members. Reports related to *Alicyclobacillus* are often presented at the annual symposium held every year. The IFU Working Group Microbiology (formerly Chairman, Mr. Bob Hartog) has been working hard on the standardization of the test method for *Alicyclobacillus* and results are now being seen. At present, Japan is not an official member of IFU but joins the Working Group Microbiology as a working group member. IFU is pouring its energies into international harmonization and dissemenating knowledge on *Alicyclobacillus*.

9-3-2 TNO Nutrition and Food Research

Homepage address: http://www.tno.nl/

TNO (Netherlands Organization for Applied Scientific Research), established in 1932, is the largest research and technology organization in Europe, where about 5,300 specialists utilizing the most advanced technologies, are working on a variety of different research projects. Mr. Bob Hartog formerly a member of TNO, collaborates with other staff on microbiological related research. *Alicyclobacillus* is one of their project themes and they carry out consulting services in addition to development of identification methods by molecular biological techniques and examination of control measures.

9-3-3 University of Applied Science

Dr. Jürgen Baumgart, Faculty of Food Microbiology, The University of Applied Science, Germany, carried out research and studies on the quality control of *Alicyclobacillus*; and in 1997 at FLÜSSIGENS OBST (Jg. 64) and at IFU Workshop Microbiology in 2000, proposed a detection method that could be applicable for use in a QC laboratory of a production factory.

9-4 Australia

9-4-1 Food Science Australia

Homepage address: http://www.foodscience.afisc.csiro.au/

Food Science Australia, an institution financed by CSIRO (http://www.csiro.au/) and AFISC (http://www.dpi.vic.gov.au/dpi/), conducts research and development with regard to foods. As a member of Food Science Australia (Food Safety & Hygiene) and also as a member of IFU Working Group Microbiology. Ms. Nancy Jensen has through presentations at symposiums and such contributed to making research on *Alicyclobacillus* more widely known.

9-4-2 Australian Fruit Juice Association

Homepage address: http://www.afja.com.au/

The Australian Fruit Juice Association, has presented at the IFU symposium, information on incidents in Australia and New Zealand of contamination due to *Alicyclobacillus*.

9-5 United States of America

9-5-1 National Food Producer Association (NFPA)

NFPA, an organization representative of the American food industry (now combined to International Association for Food Protection: http://www.foodprotection.org/), also actively pursues problems related to *Alicyclobacillus* (through research and investigation, designation of standards, and so on). Among such activities the study done by a group supervised by Dr. Isabel Walls (now Office of Food Defense and Emergency Response, USDA Food Safety and Inspection Service, through ILSI Risk Science Institute) "the pathogenicity of *Alicyclobacillus*" is the only report concerned with nonpathogenicity of this bacterium. Furthermore they have been active as representatives of NFPA, in for instance the standardization of test methods, and have contributed to the dissemination of information in the United States.

9-5-2 University of Florida

Homepage address: http://www.ufl.edu/

Dr. Mickey E. Parish and his colleagues of the Citrus Research & Education Center, University of Florida, are energetically taking on research involved in development of test methods and control measures for *Alicyclobacillus*. They have been spreading information by presentating research results on *Alicyclobacillus* at domestic symposiums and so on.

9-5-3 Washington State University

Homepage address: http://www.wsu.edu/

The group led by Dr. Dong-Hyun Kang, Washington State University (Food Science and Human Nutrition), and his colleagues is developing control methods for *Alicyclobacillus* and fast detection methods for guaiacol; reporting their results mainly in academic publications.

9-5-4 University of Georgia

Homepage address: http://www.uga.edu/

The research group of Dr. Larry Beuchat at the Center for Food Safety and Quality Enhancement, the University of Georgia works on the relationships between *Alicyslobacillus* and guaiacol, and is developing control methods for them. They have reported their results mainly in academic publications.

9-5-5 University of Wisconsin

Homepage address: http://www.wisc.edu/

Dr. Charles W. Kaspar, Food Research Institute, University of Wisconsin, and his colleagues are conducting research on *Alicyclobacillus* as part of their research on microbial control (FRI News Letter, Fall 2000, Vol. 12, No. 3).

9-5-6 Technical Committee for Juice and Juice Products (TCJJP)

Homepage address: http://www.tcjjp.org/

TCJJP is an NPO that was established to contribute to the fields related to juice, and it makes an effort to develop official test methods for bacteria including *Alicyclobacillus* in cooperation with AOAC, NJPA (http://www.njpa.com/), and NFPA.

9-5-7 Cornell University

Homepage address: http://www.cornell.edu/

Dr. Don F. Splittstoesser (deceased), an emeritus professor of Cornell University, was involved in working on the problems related to *Alicyclobacillus* and made available invaluable information.

9-5-8 International Society of Beverage Technologists (ISBT)

Homepage address: http://www.bevtech.org/

In April, 2005, an *Alicyclobacillus* (ACB) subcommittee was formed within the Quality Technical Committee of the International Society of Beverage Technology (ISBT). The subcommittee is composed of three working groups focused on Research, Methodology and Control. The Research working group is reviewing published literature to assess the state-of-the-art, and provides on-going updates to the Methodology and Control groups, and recommends consultants/subject matter experts. The Methodology group is reviewing and assessing the development of conventional and rapid methodologies, and has created a reference strain collection (of naturally-occurring strains) and is working on the publication of standardized test methods. The Control group is investigating ingredient/product process flows and sanitizer efficacy with the aim at publishing industry standards/guidelines and best practices.

Kanjiro Takahashi, Keiichi Goto

Index

MIX
Papier aus verantwortungsvollen Quellen
Paper from responsible sources
FSC® C105338
FSC
www.fsc.org

If you have any concerns about our products,
you can contact us on
ProductSafety@springernature.com

In case Publisher is established outside the EU,
the EU authorized representative is:
Springer Nature Customer Service Center GmbH
Europaplatz 3, 69115 Heidelberg, Germany

Printed by Libri Plureos GmbH
in Hamburg, Germany